Method of
Lines Analysis
of Turing Models

Method of Lines Analysis of Turing Models

William E Schiesser

Lehigh University, USA

 **World Scientific**

NEW JERSEY · LONDON · SINGAPORE · BEIJING · SHANGHAI · HONG KONG · TAIPEI · CHENNAI · TOKYO

Published by

World Scientific Publishing Co. Pte. Ltd.
5 Toh Tuck Link, Singapore 596224
USA office: 27 Warren Street, Suite 401-402, Hackensack, NJ 07601
UK office: 57 Shelton Street, Covent Garden, London WC2H 9HE

Library of Congress Cataloging-in-Publication Data
Names: Schiesser, W. E., author.
Title: Method of lines analysis of Turing models / by W.E. Schiesser (Lehigh University, USA).
Description: New Jersey : World Scientific, [2017] |
    Includes bibliographical references and index.
Identifiers: LCCN 2017023364 | ISBN 9789813226692 (hardcover : alk. paper)
Subjects: LCSH: Differential equations, Partial--Mathematical models.
Classification: LCC QA377 .S3539 2017 | DDC 515/.353--dc23
LC record available at https://lccn.loc.gov/2017023364

British Library Cataloguing-in-Publication Data
A catalogue record for this book is available from the British Library.

Copyright © 2017 by World Scientific Publishing Co. Pte. Ltd.

All rights reserved. This book, or parts thereof, may not be reproduced in any form or by any means, electronic or mechanical, including photocopying, recording or any information storage and retrieval system now known or to be invented, without written permission from the publisher.

For photocopying of material in this volume, please pay a copying fee through the Copyright Clearance Center, Inc., 222 Rosewood Drive, Danvers, MA 01923, USA. In this case permission to photocopy is not required from the publisher.

Typeset by Stallion Press
Email: enquiries@stallionpress.com

Printed in Singapore

Contents

Preface

This book is directed toward the numerical integration (solution) of a system of partial differential equations (PDEs) that describe a combination of chemical reaction and diffusion, that is, *reaction-diffusion* PDEs. The particular form of the PDEs corresponds to a system discussed by Turing [1] and is therefore termed a *Turing model*.

Specifically, Turing considerd how a reaction-diffusion system can be formulated that does not have the usual smoothing properties of a diffusion (dispersion) system, and can, in fact, develop a spatial pattern (variation) that might be interpreted as a form of *morphogensis*, so he termed the chemicals as *morphogens*.

Turing alluded to the important impact computers would have in the study of a morphogenic PDE system, but at the time (1952), computers were still not readily available[1]. Therefore, his paper is based on analytical methods. Although computers have since been applied to Turing models, to the author's knowledge, computer-based analysis is still not facilitated by a discussion

---

[1]Turing concluded his paper with a section titled *Non-linear theory. Use of digital computers* in which he anticipated the use of numerical methods for the analysis of nonlinear PDEs. This book is an attempt to confirm his expectation of the use of computers for the analysis of reaction-diffusion PDEs as applied to biological systems.

of numerical algorithms and a readily available system (set) of computer routines.

The intent of this book is to provide a basic discussion of numerical methods and associated computer routines for reaction-diffusion systems of varying form. The presentation has a minimum of formal mathematics. Rather, the presentation is in terms of detailed examples, presented at an introductory level. This format should assist readers who are interested in developing computer-based analysis for reaction-diffusion PDE systems without having to first study numerical methods and computer programming (coding).

The first examples are for 1D PDEs, with linear and non-linear reaction terms. The numerical examples are discussed in terms of: (1) numerical integration of the PDEs to demonstrate the spatiotemporal features (patterns) of the solutions and (2) a numerical eigenvalue analysis that corroborates the observed temporal variation of the solutions.

The examples are then extended to 2D with plotting of the solutions in 3D perspective. The resulting temporal variation of the 3D plots demonstrates how the solutions evolve dynamically, including stable or unstable long-term behavior.

In all of the examples, routines in R^2 are presented and discussed in detail. The routines are available through a download link so that the reader can execute the PDE models to reproduce the reported solutions, then experiment with the models, including extensions and application to alternative models.

In summary, the presentation is not as formal mathematics, e.g., theorems and proofs. Rather, the presentation is by

---

[2]R is an open-source scientific programming system that is easily downloaded from the Internet. It has utilities for linear algebra, numerical integration and graphical output that facilitate the study of reaction-diffusion PDE models.

introductory examples of reaction-diffusion PDE models, including the details for computing numerical solutions, particularly with documented, transportable source code. The author would welcome comments and suggestions for improvements (wes1@lehigh.edu).

William E. Schiesser
Bethlehem, PA, USA
February 1, 2017

Reference

[1] Turing, A.M. (1952), The chemical basis of morphogenesis, *Philosophical Transactions of the Royal Society of London, Series B, Biological Sciences*, **237**, no. 641, 37–72

Chapter 1

One Dimensional PDEs

Introduction

Reaction-diffusion (RD) equations are an important class of partial differential equations (PDEs) with broad application in physics, chemistry, biology. Because of this general applicability, they have received extensive discussion.

Various forms of RD PDEs are introduced in this chapter. Numerical methods and associated computer analysis of specific PDE systems are then considered in this and subsequent chapters. The discussion is limited to 2×2 (two equations in two unknowns) and 3×3 PDE systems, but extensions to larger PDE systems is straightforward.

(1.1) Various Coordinate Systems

The following 3×3 RD system in coordinate-free form is the starting point for the subsequent discussion of PDEs in specific coordinate systems.

$$\frac{\partial u_1}{\partial t} = r_1(u_1, u_2, u_3) + D_1 \nabla^2 u_1 \qquad (1.1a)$$

$$\frac{\partial u_2}{\partial t} = r_2(u_1, u_2, u_3) + D_2 \nabla^2 u_2 \qquad (1.1b)$$

$$\frac{\partial u_3}{\partial t} = r_3(u_1, u_2, u_3) + D_3 \nabla^2 u_3 \qquad (1.1c)$$

1

where

u_1, u_2, u_3 dependent variables (concentrations)

t initial value independent variable, typically time

$\nabla^2 = \nabla \bullet \nabla$ Laplacian operator

$\bullet$ vector product

∇ gradient of a scalar

$\nabla \bullet$ divergence of a vector

r_1, r_2, r_3 volumetric rates of reaction of u_1, u_2, u_3, respectively

Eqs. (1.1) indicate that the coupling between the equations is through the reaction terms r_1, r_2, r_3. Note that the diffusion terms are linear[1].

---

[1]Eqs. (1.1) are based on the diffusion flux, $\mathbf{q}$,

$$\mathbf{q} = -D\nabla u$$

that is, a linear function of the gradient ∇u which is termed *Fick's first law*. Departures from this linear form can be analyzed numerically, including the use of the routines that are discussed subsequently.

Eqs. (1.1) are derived from mass balances ([1], Sec. A1.4). Eqs. (1.1) in Cartesian coordinates x, y, z are

$$\frac{\partial u_1}{\partial t} = r_1(u_1, u_2, u_3) + D_1 \left(\frac{\partial^2 u_1}{\partial x^2} + \frac{\partial^2 u_1}{\partial y^2} + \frac{\partial^2 u_1}{\partial z^2} \right) \qquad (1.2a)$$

$$\frac{\partial u_1}{\partial t} = r_2(u_1, u_2, u_3) + D_2 \left(\frac{\partial^2 u_2}{\partial x^2} + \frac{\partial^2 u_2}{\partial y^2} + \frac{\partial^2 u_2}{\partial z^2} \right) \qquad (1.2b)$$

$$\frac{\partial u_1}{\partial t} = r_3(u_1, u_2, u_3) + D_3 \left(\frac{\partial^2 u_3}{\partial x^2} + \frac{\partial^2 u_3}{\partial y^2} + \frac{\partial^2 u_3}{\partial z^2} \right) \qquad (1.2c)$$

Eqs. (1.2) are *parabolic* PDEs with reaction (first order in t and second order in x, y, z), This contrasts with *hyperbolic* PDEs (first order in x and t).

In the subsequent analysis, the 1D special case of eqs. (1.2) is considered (the derivatives in y, z are dropped)[2]. This 1D system is then the basis for the RD PDEs considered by Turing [2].

---

[2]Eq. (1.1a) in cylindrical coordinates r, θ, z follows.

$$\frac{\partial u_1}{\partial t} = r_1(u_1, u_2, u_3) + D_1 \left(\frac{\partial^2 u_1}{\partial r^2} + \frac{1}{r}\frac{\partial u_1}{\partial r} + \frac{1}{r^2}\frac{\partial^2 u_1}{\partial \theta^2} + \frac{\partial^2 u_1}{\partial z^2} \right)$$
$$(1.2d)$$

Similar equations follow for eqs. (1.1b), (1.1c).

Eq. (1.1a) in spherical coordinates r, θ, ϕ follows.

$$\frac{\partial u_1}{\partial t} = r_1(u_1, u_2, u_3) + D_1 \left(\frac{\partial^2 u_1}{\partial r^2} + \frac{2}{r}\frac{\partial u_1}{\partial r} \right.$$

$$\left. + \frac{1}{r^2} \left(\frac{\partial^2 u_1}{\partial \theta^2} + \frac{\cos(\theta)}{\sin(\theta)}\frac{\partial u_1}{\partial \theta} \right) + \frac{1}{r^2 \sin^2(\theta)}\frac{\partial^2 u_1}{\partial \phi^2} \right) \qquad (1.2e)$$

(1.2) No Reaction Model

For $r_1(u_1, u_2, u_3) = r_2(u_1, u_2, u_3) = r_3(u_1, u_2, u_3) = 0$, eqs. (1.1) reduce to

$$\frac{\partial u_i}{\partial t} = D_i \nabla^2 u_i \qquad (1.3)$$

$i = 1, 2, 3$, Eqs. (1.3) are a 3×3 system of uncoupled PDEs. In fact, each PDE is just the *diffusion equation*, also termed *Fick's second law*. This special case is used subsequently to test some of the numerical methods and coding.

(1.3) No Diffusion Model

For $D_1 = D_2 = D_3 = 0$, eqs. (1.1) reduce to

$$\frac{du_1}{dt} = r_1(u_1, u_2, u_3) \qquad (1.4a)$$

$$\frac{du_2}{dt} = r_2(u_1, u_2, u_3) \qquad (1.4b)$$

$$\frac{du_3}{dt} = r_3(u_1, u_2, u_3) \qquad (1.4c)$$

Eqs. (1.4) are a system of ordinary differential equations (ODEs) with no spatial effects, termed a *discrete* or *compartment* model. There can be interconnected compartments, each perfectly mixed so that there are no spatial effects within each compartment.

(1.4) Linear Reaction-diffusion Model

If the reactions rates r_1, r_2, r_3 are linear functions of the dependent variables, eqs. (1.1) in 1D are

$$\frac{\partial u_1}{\partial t} = a_{11}u_1 + a_{12}u_2 + a_{13}u_3 + D_1\frac{\partial^2 u_1}{\partial x^2} \qquad (1.5a)$$

$$\frac{\partial u_2}{\partial t} = a_{21}u_1 + a_{22}u_2 + a_{23}u_3 + D_2\frac{\partial^2 u_2}{\partial x^2} \qquad (1.5b)$$

$$\frac{\partial u_3}{\partial t} = a_{31}u_1 + a_{32}u_2 + a_{33}u_3 + D_3\frac{\partial^2 u_3}{\partial x^2} \qquad (1.5c)$$

where $a_{11}, \ldots, a_{33}$ are constants to be specified. Eqs. (1.5) are linear, constant coefficient PDEs that can be analyzed by the usual methods of linear algebra if the diffusion terms are replaced by linear approximations such as finite differences (FDs). The resulting system of ODEs in t is the starting point for the *method of lines* (MOL)[3]. This MOL based on FDs is considered in [2] and is termed a *Turing model*.

(1.5) Computer Routines

The computer analysis of the preceding PDE systems is implemented with a series of routines that are next listed and discussed in detail.

(1.5.1) Main program

A main program for eqs. (1.5) follows.

```
#
# 1D 3 x 3 Turing
#
# Delete previous workspaces
  rm(list=ls(all=TRUE))
#
```

---

[3]The MOL is a general numerical method for PDEs in which the spatial derivatives are replaced by algebraic approximations, e.g., FDs, finite volumes (FVs), finite elements (FEs), least squares, spectral approximations. The resulting system of ODEs in an initial value variable, usually time, is integrated numerically, typically by a library ODE integrator.

```
# Access ODE integrator
  library("deSolve");
#
# Access functions for numerical solution
  setwd("f:/turing/1D");
  source("pde_1a.R");
#
# Grid in x
  xl=0;xu=1;n=51;dx=(xu-xl)/(n-1);
  x=seq(from=xl,to=xu,by=dx);
  n2=2*n;n3=3*n;
#
# Parameters
  ncase=1;
  if(ncase==1){
    a11  =0; a12  =0; a13  =0;
    a21  =0; a22  =0; a23  =0;
    a31  =0; a32  =0; a33  =0;
    D1=dx^2; D2=dx^2; D3=dx^2;
    t0   =0; tf   =20; nout =6;
    c1   =0; c2   =50;
    zl   =0; zu   =1;}
  if(ncase==2){
    a11= -1; a12  =0; a13  =0;
    a21 = 0; a22 =-1; a23 = 0;
    a31 = 0; a32  =0; a33 =-1;
    D1=dx^2; D2=dx^2; D3=dx^2;
    t0   =0; tf   =20; nout =6;
    c1   =0; c2   =50;
    zl   =0; zu   =1;}
  if(ncase==3){
    a11  =1; a12  =0; a13  =0;
    a21  =0; a22  =1; a23  =0;
    a31  =0; a32  =0; a33  =1;
```

```
  D1=dx^2;  D2=dx^2;  D3=dx^2;
  t0   =0; tf    =2; nout =6;
  c1   =0; c2   =50;
  zl   =0; zu    =1;}
if(ncase==4){
a11  =-10/3; a12      =3; a13    =-1;
a21       =-2; a22    =7/3; a23    =0;
a31       =3; a32      =-4; a33    =0;
D1=2/3*dx^2;  D2=1/3*dx^2;  D3=0*dx^2;
t0        =0; tf      =20; nout   =6;
c1        =0; c2      =50;
zl        =-2; zu      =3;}
if(ncase==5){
a11   =-1; a12   =-1; a13   = 0;
a21   =1; a22   =0; a23   =-1;
a31   =0; a32   =1; a33   =0;
D1=1*dx^2;  D2=0*dx^2;  D3=0*dx^2;
t0    =0; tf    =20; nout   =6;
c1    =0; c2    =50;
zl    =-1; zu    =2;}
#
# Factors used in pde_1a.R
  D1dx2=D1/dx^2;D2dx2=D2/dx^2;D3dx2=D3/dx^2;
#
# Independent variable for ODE integration
  tout=seq(from=t0,to=tf,by=(tf-t0)/(nout-1));
#
# ICs
  u0=rep(0,n3);
  for(i in 1:n){
    u0[i]   =c1+exp(-c2*(x[i]-0.5)^2);
    u0[i+n] =c1+exp(-c2*(x[i]-0.5)^2);
    u0[i+n2]=c1+exp(-c2*(x[i]-0.5)^2);
#    u0[i+n] =0;
```

```
#   u0[i+n2]=0;
  }
  ncall=0;
#
# ODE integration
  out=lsodes(y=u0,times=tout,func=pde_1a,
      sparsetype ="sparseint",rtol=1e-6,
      atol=1e-6,maxord=5);
  nrow(out)
  ncol(out)
#
# Arrays for numerical solution
  u1=matrix(0,nrow=n,ncol=nout);
  u2=matrix(0,nrow=n,ncol=nout);
  u3=matrix(0,nrow=n,ncol=nout);
  t=rep(0,nout);
  for(it in 1:nout){
  for(i  in 1:n){
    u1[i,it]=out[it,i+1];
    u2[i,it]=out[it,i+1+n];
    u3[i,it]=out[it,i+1+n2];
      t[it]=out[it,1];
  }
  }
#
# Display selected output
  for(it in 1:nout){
    cat(sprintf("\n      t        x    u1(x,t)
            u2(x,t)    u3(x,t)\n"));
    iv=seq(from=1,to=n,by=5);
    for(i in iv){
      cat(sprintf(
        "%6.2f%9.3f%10.6f%10.6f%10.6f\n",
        t[it],x[i],u1[i,it],u2[i,it],u3[i,it]));
```

```
    }
    cat(sprintf("\n"));
  }
  cat(sprintf(" ncall = %4d\n",ncall));
#
# Plot 2D numerical solution
  matplot(x,u1,type="l",lwd=2,col="black",
    lty=1,xlab="x",ylab="u1(x,t)",main="");
  matplot(x,u2,type="l",lwd=2,col="black",
    lty=1,xlab="x",ylab="u2(x,t)",main="");
  matplot(x,u3,type="l",lwd=2,col="black",
    lty=1,xlab="x",ylab="u3(x,t)",main="");
#
# Plot 3D numerical solution
  persp(x,t,u1,theta=45,phi=45,xlim=c(xl,xu),
        ylim=c(t0,tf),xlab="x",ylab="t",
        zlab="u1(x,t)");
  persp(x,t,u2,theta=45,phi=45,xlim=c(xl,xu),
        ylim=c(t0,tf),xlab="x",ylab="t",
        zlab="u2(x,t)");
  persp(x,t,u3,theta=45,phi=45,xlim=c(xl,xu),
        ylim=c(t0,tf),xlab="x",ylab="t",
        zlab="u3(x,t)");
```

Listing 1.1: Main program for eqs. (1.5)

We can note the following details about Listing 1.1.

- Previous workspaces are removed. Then the ODE integrator library **deSolve** is accessed. Note that the **setwd** (set working directory) uses / rather than the usual \.

  ```
  #
  # 1D 3 x 3 Turing
  #
  # Delete previous workspaces
  ```

```
   rm(list=ls(all=TRUE))
#
# Access ODE integrator
   library("deSolve");
#
# Access functions for numerical solution
   setwd("f:/turing/1D");
   source("pde_1a.R");
```

pde_1a is the routine for the MOL ODEs (discussed subsequently).

- A uniform grid in x of 51 points is defined with the seq utility for the interval $x_l = 0 \leq x \leq x_u = 1$. Therefore, the vector x has the values $x = 0, 1/50, \ldots, 1$.

```
#
# Grid in x
   xl=0;xu=1;n=51;dx=(xu-xl)/(n-1);
   x=seq(from=xl,to=xu,by=dx);
   n2=2*n;n3=3*n;
```

- Five cases are programmed with variations in the model parameters

```
#
# Parameters
   ncase=1;
   if(ncase==1){
      a11  =0; a12  =0; a13  =0;
      a21  =0; a22  =0; a23  =0;
      a31  =0; a32  =0; a33  =0;
      D1=dx^2; D2=dx^2; D3=dx^2;
      t0   =0; tf   =20; nout =6;
      c1   =0; c2   =50;
      zl   =0; zu   =1;}
```

```
if(ncase==5){
a11    =-1; a12    =-1; a13    = 0;
a21    =1; a22     =0; a23    =-1;
a31    =0; a32     =1; a33    =0;
D1=1*dx^2; D2=0*dx^2; D3=0*dx^2;
t0     =0; tf      =20; nout   =6;
c1     =0; c2      =50;
zl     =-1; zu     =2;}
```

Briefly, these cases are:

- **ncase==1**: The no reaction eqs. (1.3) are programmed (all of the coefficients in the reaction terms are zero). The diffusivities are defined as the square of the grid spacing in x, dx^2, in accordance with [2] (this will be explained further when the ODE/MOL routine in Listing 1.2 is discussed). The interval in t is $t_0 \leq t \leq t_f$ (discussed next). The constants in the Gaussian IC, $c_1 + e^{-c_2(x-0.5)^2}$, are defined (discussed subsequently). The z interval in the 3D plots is defined as $z_l \leq z \leq z_u$ (discsussed subsequently). The numerical solution is therefore for the diffusion eq. (1.3), repeated three times (to give a smooth decay in t).
- **ncase==2**: ncase==1 with diagonal reaction terms added, a11=a22=a33=-1. These negative values correspond to consumption of the reactants and therefore the solution is stable as in **ncase==1** (a smooth decay in t that is faster than for **ncase==1** since the reaction and diffusion are superimposed).
- **ncase==3**: ncase==1 with diagonal reaction terms added, a11=a22=a33=1. These positive values

correspond to production of the reactants and therefore the solution is unstable (increases in t without bound).

– ncase==4: The reaction constants $a_{11}, \ldots, a_{33}$ and diffusivities are taken from Turing ([2], p53, eqs. (8.5)).

– ncase==5: The reaction constants $a_{11}, \ldots, a_{33}$ and diffusivities are taken from Turing ([2], p54, eqs. (8.7)).

For ncase==1,2,3, the eigenvalues λ_i of the approximating MOL/ODEs are real with $Re\lambda_i \leq 0$ for a stable solution (ncase=1,2), and (at least one) $Re\lambda_i > 0$ for an unstable solution (ncase==3). A central idea by Turing is that the reaction coefficients $a_{11}, \ldots, a_{33}$ and diffusivities D_1, D_2, D_3 are selected so that some of the eigenvalues are complex (conjugate pairs) and the solution therefore oscillates in t. This is a marked distinction from Fickian diffusion which disperses the solution monotonically in t. This diffusion with oscillation is discussed further in Chapter 2.

- Factors used in the ODE/MOL routine pde_1a are defined.

```
#
# Factors used in pde_1a.R
  D1dx2=D1/dx^2;D2dx2=D2/dx^2;D3dx2=D3/dx^2;
```

Each term has a division by dx^2 which cancels the same factor in the definition of D1,D2,D3 so that these diffusivities correspond to their use by Turing ([2], p47, eq. 6.2 applied to three components u_1, u_2, u_3).

- A uniform grid in t of 6 output points is defined with the seq utility for the interval $t_0 = 0 \leq t \leq t_f = 20$. Therefore, the vector tout has the values $t = 0, 4, \ldots, 20$.

`ncase==3` is an exception with $t_f = 2$ because of the rapid approach to instability in t.

```
#
# Independent variable for ODE integration
  tout=seq(from=t0,to=tf,by=(tf-t0)/(nout-1));
```

At this point, the intervals of x and t in eq. (1.5) are defined.

- IC functions for eqs. (1.5) are defined. `n2,n3` are constants defined previously.

```
#
# ICs
  u0=rep(0,n3);
  for(i in 1:n){
    u0[i]   =c1+exp(-c2*(x[i]-0.5)^2);
    u0[i+n] =c1+exp(-c2*(x[i]-0.5)^2);
    u0[i+n2]=c1+exp(-c2*(x[i]-0.5)^2);
#   u0[i+n] =0;
#   u0[i+n2]=0;
  }
  ncall=0;
```

The functions for u_2, u_3 can also be activated (uncommented) to zero to study the transfer from u_1 to u_2, u_3. Another possibility would be to use randomly distributed IC values to investigate patterning in the solutions, but this might require an increase in the number of grid points in x (above `n=51`) for improved spatial resolution. The counter for the calls to `pde_1a` is initialized (and passed to `pde_1a` without a special designation).

- The system of $3(51) = 153$ MOL/ODEs is integrated by the library integrator `lsodes` (available in `deSolve`) with the sparse matrix option specified. As expected, the inputs to `lsodes` are the ODE function, `pde_1a`, the IC vector u0, and the vector of output values of t, tout. The length of u0 (e.g., 153) informs lsodes how many ODEs are to be integrated. func,y,times are reserved names.

```
#
# ODE integration
  out=lsodes(y=u0,times=tout,func=pde_1a,
      sparsetype ="sparseint",rtol=1e-6,
      atol=1e-6,maxord=5);
  nrow(out)
  ncol(out)
```

The numerical solution to the ODEs is returned in matrix out. In this case, out has the dimensions $nout \times (n+1) = 6 \times 154$. The offset $n+1$ is required since the first element of each column has the output t (also in tout), and the $2, \ldots, n+1 = 2, \ldots, 154$ column elements have the 153 ODE solutions. This indexing of out in used next.

• The ODE solution is placed in 3 51×6 matrices, u1,u2,u3, for subsequent plotting (by stepping through the solution with respect to x and t within a pair of fors).

```
#
# Arrays for numerical solution
  u1=matrix(0,nrow=n,ncol=nout);
  u2=matrix(0,nrow=n,ncol=nout);
  u3=matrix(0,nrow=n,ncol=nout);
  t=rep(0,nout);
  for(it in 1:nout){
  for(i  in 1:n){
    u1[i,it]=out[it,i+1];
    u2[i,it]=out[it,i+1+n];
    u3[i,it]=out[it,i+1+n2];
      t[it]=out[it,1];
  }
  }
```

- The numerical solutions are displayed.

```
#
# Display selected output
  for(it in 1:nout){
    cat(sprintf("\n      t          x    u1(x,t)
          u2(x,t)    u3(x,t)\n"));
    iv=seq(from=1,to=n,by=5);
    for(i in iv){
      cat(sprintf(
        "%6.2f%9.3f%10.6f%10.6f%10.6f\n",
        t[it],x[i],u1[i,it],u2[i,it],u3[i,it]));
      }
    cat(sprintf("\n"));
    }
  cat(sprintf(" ncall = %4d\n",ncall));
```

To conserve space, only every fifth value in x of the solutions is displayed numerically (using the subscript `iv`).
- The solutions to eqs. (1.5), $u_1(x,t), u_2(x,t), u_3(x,t)$, are plotted vs x with t as a parameter in 2D with `matplot`.

```
#
# Plot 2D numerical solution
  matplot(x,u1,type="l",lwd=2,col="black",
    lty=1,xlab="x",ylab="u1(x,t)",main="");
  matplot(x,u2,type="l",lwd=2,col="black",
    lty=1,xlab="x",ylab="u2(x,t)",main="");
  matplot(x,u3,type="l",lwd=2,col="black",
    lty=1,xlab="x",ylab="u3(x,t)",main="");
```

Note that the rows of rows of `x` (`nrows=n=51`) equals the rows of `u1,u2,u3`.
- The solutions to eqs. (1.5), $u_1(x,t), u_2(x,t), u_3(x,t)$, are plotted vs x and t in 3D perspective with `persp`.

```
#
# Plot 3D numerical solution
  persp(x,t,u1,theta=45,phi=45,xlim=c(xl,xu),
        ylim=c(t0,tf),xlab="x",ylab="t",
        zlab="u1(x,t)");
  persp(x,t,u2,theta=45,phi=45,xlim=c(xl,xu),
        ylim=c(t0,tf),xlab="x",ylab="t",
        zlab="u2(x,t)");
  persp(x,t,u3,theta=45,phi=45,xlim=c(xl,xu),
        ylim=c(t0,tf),xlab="x",ylab="t",
        zlab="u3(x,t)");
```

Automatic scaling in `z` is used. If the three plots are to have a common vertical scale in `z`, `zlim=c(zl,zu)` defined previously could be used. `x,t` have dimensions in agreement with `u1,u2,u3` (`nrows=n=51`, `ncols=nout=6`).

The ODE/MOL routine called in the main program of Listing 1.1, `pde_1a`, follows.

(1.5.2) ODE/MOL routine

The ODE/MOL routine for eqs. (1.5) is listed next.

```
  pde_1a=function(t,u,parm){
#
# Function pde_1a computes the t derivative
# vector for u1(x,t), u2(x,t), u3(x,t)
#
# One vector to three vectors
  u1 =rep(0,n);u2 =rep(0,n);u3 =rep(0,n);
  u1t=rep(0,n);u2t=rep(0,n);u3t=rep(0,n);
  for(i in 1:n){
    u1[i]=u[i];
```

```
    u2[i]=u[i+n];
    u3[i]=u[i+n2];
  }
#
# u1t(x,t)
  for(i in 1:n){
    if(i==1){u1t[1]=a11*u1[1]+a12*u2[1]+a13*u3[1]+
            2*D1dx2*(u1[  2]-u1[1]);}
    if(i==n){u1t[n]=a11*u1[n]+a12*u2[n]+a13*u3[n]+
            2*D1dx2*(u1[n-1]-u1[n]);}
    if((i>1)&(i<n)){
      u1t[i]=a11*u1[i]+a12*u2[i]+a13*u3[i]+
            D1dx2*(u1[i+1]-2*u1[i]+u1[i-1]);}
  }
#
# u2t(x,t)
  for(i in 1:n){
    if(i==1){u2t[1]=a21*u1[1]+a22*u2[1]+a23*u3[1]+
            2*D1dx2*(u2[  2]-u2[1]);}
    if(i==n){u2t[n]=a21*u1[n]+a22*u2[n]+a23*u3[n]+
            2*D1dx2*(u2[n-1]-u2[n]);}
    if((i>1)&(i<n)){
      u2t[i]=a21*u1[i]+a22*u2[i]+a23*u3[i]+
            D1dx2*(u2[i+1]-2*u2[i]+u2[i-1]);}
  }
#
# u3t(x,t)
  for(i in 1:n){
    if(i==1){u3t[1]=a31*u1[1]+a32*u2[1]+a33*u3[1]+
            2*D3dx2*(u3[  2]-u3[1]);}
    if(i==n){u3t[n]=a31*u1[n]+a32*u2[n]+a33*u3[n]+
            2*D3dx2*(u3[n-1]-u3[n]);}
    if((i>1)&(i<n)){
      u3t[i]=a31*u1[i]+a32*u2[i]+a33*u3[i]+
```

```
                      D3dx2*(u3[i+1]-2*u3[i]+u3[i-1]);}
  }
#
# Three vectors to one vector
  ut=rep(0,n3);
  for(i in 1:n){
    ut[i]    =u1t[i];
    ut[i+n]  =u2t[i];
    ut[i+n2]=u3t[i];
  }
#
# Increment calls to pde_1a
  ncall <<- ncall+1;
#
# Return derivative vector
  return(list(c(ut)));
  }
```

Listing 1.2: ODE/MOL routine `pde_1a` for eqs. (1.5)

We can note the following details about `pde_1a`.

- The function is defined.

```
    pde_1a=function(t,u,parms){
  #
  # Function pde_1a computes the t derivative
  # vector for u1(x,t), u2(x,t), u3(x,t)
```

  t is the current value of t in eqs. (1.5). u is the 153-vector
  of ODE/MOL dependent variables. `parm` is an argument
  to pass parameters to `pde_1a` (unused, but required in the
  argument list). The arguments must be listed in the order
  stated to properly interface with `lsodes` called in the

main program of Listing 1.1. The composite derivative vector of the LHSs of eqs. (1.5) is calculated next and returned to lsodes.

- The dependent variable vectors are placed in three vectors to facilitate the programming of eqs. (1.5).

```
#
# One vector to three vectors
  u1 =rep(0,n);u2 =rep(0,n);u3 =rep(0,n);
  u1t=rep(0,n);u2t=rep(0,n);u3t=rep(0,n);
  for(i in 1:n){
    u1[i]=u[i];
    u2[i]=u[i+n];
    u3[i]=u[i+n2];
  }
```

Vectors are also defined for the LHS derivatives in t of eqs. (1.5).

- $\dfrac{\partial u_1}{\partial t}$ of eq. (1.5a) is programmed in a for that steps through x, for(i in 1:n).

```
#
# u1t(x,t)
  for(i in 1:n){
    if(i==1)
      {u1t[1]=a11*u1[1]+a12*u2[1]+a13*u3[1]+
            2*D1dx2*(u1[  2]-u1[1]);}
    if(i==n)
      {u1t[n]=a11*u1[n]+a12*u2[n]+a13*u3[n]+
            2*D1dx2*(u1[n-1]-u1[n]);}
    if((i>1)&(i<n)){
      u1t[i]=a11*u1[i]+a12*u2[i]+a13*u3[i]+
          D1dx2*(u1[i+1]-2*u1[i]+u1[i-1]);}
  }
```

This coding requires some additional explanation.

— At $x = x_l = 0(i = 1)$, the homogeneous Neumann BC $\dfrac{\partial u_1(x = 0, t)}{\partial x} = 0$ is used. For a FD MOL with grid spacing Δx, the approximation of the BC is

$$\frac{\partial u_1(x = 0, t)}{\partial x} \approx$$

$$\frac{u_1(x = \Delta x, t) - u_1(x = -\Delta x, t)}{2\Delta x} = 0$$

The fictitious value is therefore $u_1(x = -\Delta x, t) = u_1(x = \Delta x, t)$ which is used in the FD approximation of $\dfrac{\partial^2 u_1(x = 0, t)}{\partial x^2}$ of eq. (1.5a)[4],

$$\frac{\partial^2 u_1(x = 0, t)}{\partial x^2} \approx$$

$$\frac{u_1(x = \Delta x, t) - 2u_1(x = 0, t) + u_1(x = -\Delta x, t)}{\Delta x^2}$$

$$= 2\frac{u_1(x = \Delta x, t) - u_1(x = 0, t)}{\Delta x^2}$$

or 2*D1dx2*(u1[2]-u1[1]).
Thus, the coding for $u_1(x = 0, t)$ is

```
if(i==1)
   {u1t[1]=a11*u1[1]+a12*u2[1]+a13*u3[1]+
        2*D1dx2*(u1[ 2]-u1[1]);}
```

u1t[1]=a11*u1[1]+a12*u2[1]+a13*u3[1]   is the linear reaction term of eq. (1.5a).

---

[4]The approximation for the second derivative is discussed further in Appendices A1 and A2.

– The same reasoning applies to the application of a homogeneous Neumann BC at the right boundary $x = x_u = 1$. The final result is

$$\frac{\partial^2 u_1(x = x_u, t)}{\partial x^2} \approx$$

$$\frac{u_1(x = x_u + \Delta x, t)}{\Delta x^2}$$

$$-\frac{2u_1(x = x_u, t)}{\Delta x^2}$$

$$+\frac{u_1(x = x_u - \Delta x, t)}{\Delta x^2}$$

$$= 2\frac{u_1(x = x_u - \Delta x, t) - u_1(x = x_u, t)}{\Delta x^2}$$

or `2*D1dx2*(u1[n-1]-u1[n])`. The coding for the right boundary is therefore

```
if(i==n)
   {u1t[n]=a11*u1[n]+a12*u2[n]+a13*u3[n]+
        2*D1dx2*(u1[n-1]-u1[n]));}
```

– For the interior points $x_l + \Delta x \le x \le x_u - \Delta x$, the coding is

```
if((i>1)&(i<n)){
    u1t[i]=a11*u1[i]+a12*u2[i]+a13*u3[i]+
        D1dx2*(u1[i+1]-2*u1[i]+u1[i-1]);}
}
```

The final `}` concludes the `for` in x. This coding follows directly from [2], p47, eq. (6.2), with $1/\Delta x^2$ included in the diffusivity, that is, in the factor `D1dx2` (see Listing 1.1). In other words, the diffusivities in [2] are considered to include $1/\Delta x^2$.

- Similar coding applied to $u_2(x,t)$ (with homogeneous Neumann BCs).

```
#
# u2t(x,t)
  for(i in 1:n){
    if(i==1)
      {u2t[1]=a21*u1[1]+a22*u2[1]+a23*u3[1]+
            2*D1dx2*(u2[  2]-u2[1]);}
    if(i==n)
      {u2t[n]=a21*u1[n]+a22*u2[n]+a23*u3[n]+
            2*D1dx2*(u2[n-1]-u2[n]);}
    if((i>1)&(i<n)){
      u2t[i]=a21*u1[i]+a22*u2[i]+a23*u3[i]+
            D2dx2*(u2[i+1]-2*u2[i]+u2[i-1]);}
  }
```

- The coding for $u_3(x,t)$ is (with homogeneous Neumann BCs)

```
#
# u3t(x,t)
  for(i in 1:n){
    if(i==1)
      {u3t[1]=a31*u1[1]+a32*u2[1]+a33*u3[1]+
            2*D1dx2*(u3[  2]-u3[1]);}
    if(i==n)
      {u3t[n]=a31*u1[n]+a32*u2[n]+a33*u3[n]+
            2*D1dx2*(u3[n-1]-u3[n]);}
    if((i>1)&(i<n)){
      u3t[i]=a31*u1[i]+a32*u2[i]+a33*u3[i]+
            D3dx2*(u3[i+1]-2*u3[i]+u3[i-1]);}
  }
```

The 3×3 array consisting of the elements $a_{11}, \ldots, a_{33}$ is defined in the main program of Listing 1.1 for five cases, ncase=1,...,5.

- The three vectors u1t,u2t,u3t are placed in a single vector u of length $3(51) = 153$ for return to lsodes.

```
#
# Three vectors to one vector
  ut=rep(0,n3);
  for(i in 1:n){
    ut[i]    =u1t[i];
    ut[i+n]  =u2t[i];
    ut[i+n2]=u3t[i];
  }
```

- The counter for the calls to pde_1a is incremented and its value is returned to the calling program (of Listing 1.1) with the <<- operator.

```
#
# Increment calls to pde_1a
  ncall <<- ncall+1;
```

- The derivative vector (LHSs of eqs. (1.5)) is returned to lsodes which requires a list. c is the R vector utility. The combination of return, list, c gives lsodes (the ODE integrator called in the main program of Listing 1.5) the required derivative vector for the next step along the solution.

```
#
# Return derivative vector
  return(list(c(ut)));
}
```

The final } concludes pde_1a.

The numerical and graphical (plotted) output is considered next.

(1.6) Model Output

Abbreviated numerical output for ncase=1 follows.

[1] 6

[1] 154

t	x	u1(x,t)	u2(x,t)	u3(x,t)
0.00	0.000	0.000004	0.000004	0.000004
0.00	0.100	0.000335	0.000335	0.000335
0.00	0.200	0.011109	0.011109	0.011109
0.00	0.300	0.135335	0.135335	0.135335
0.00	0.400	0.606531	0.606531	0.606531
0.00	0.500	1.000000	1.000000	1.000000
0.00	0.600	0.606531	0.606531	0.606531
0.00	0.700	0.135335	0.135335	0.135335
0.00	0.800	0.011109	0.011109	0.011109
0.00	0.900	0.000335	0.000335	0.000335
0.00	1.000	0.000004	0.000004	0.000004

.
.
.

Output for t = 4,...,16 removed

.
.
.

t	x	u1(x,t)	u2(x,t)	u3(x,t)
20.00	0.000	0.010279	0.010279	0.010279
20.00	0.100	0.029273	0.029273	0.029273
20.00	0.200	0.109668	0.109668	0.109668
20.00	0.300	0.286933	0.286933	0.286933
20.00	0.400	0.511844	0.511844	0.511844
20.00	0.500	0.620911	0.620911	0.620911
20.00	0.600	0.511844	0.511844	0.511844

```
20.00    0.700   0.286933   0.286933   0.286933
20.00    0.800   0.109668   0.109668   0.109668
20.00    0.900   0.029273   0.029273   0.029273
20.00    1.000   0.010279   0.010279   0.010279
```

Table 1.1: Abbreviated numerical output for `ncase=1`

We can note the following details about this output.

- The values $t = 0, 4, \ldots, 20$ follow from the coding in the main program of Listing 1.1.
- The values $x = 0, 0.02, \ldots, 1$ follow from the coding in the main program of Listing 1.1, with every fifth value displayed.
- The solutions for u_1, u_2, u_3 are the same as expected for `ncase=1`. However, this check is worthwhile since different solutions would indicate an error in the coding, particularly in `pde_1a` of Listing 1.2.

Graphical output is given in Figs. 1.1a (2D) and 1.1b (3D), for $u_1(x, t)$, `ncase=1`.

Figs. 1.1 indicate a monotonic decay of the solution in t from diffusion (dispersion). Also, $n = 51$ and *nout* $= 6$ appear to give acceptable resolution in x and t.

The five cases, `ncase=1,2,3,4,5`, of Listing 1.1 produce extensive numerical and graphical output. Therefore, because of space limitations, only abbreviated output for `ncase=1,4` is considered. The preceding output is for `ncase=1` and the following output is for `ncase=4`.

[1] 6

[1] 154

```
       t        x     u1(x,t)    u2(x,t)    u3(x,t)
    0.00    0.000   0.000004   0.000004   0.000004
    0.00    0.100   0.000335   0.000335   0.000335
```

0.00	0.200	0.011109	0.011109	0.011109
0.00	0.300	0.135335	0.135335	0.135335
0.00	0.400	0.606531	0.606531	0.606531
0.00	0.500	1.000000	1.000000	1.000000
0.00	0.600	0.606531	0.606531	0.606531
0.00	0.700	0.135335	0.135335	0.135335
0.00	0.800	0.011109	0.011109	0.011109
0.00	0.900	0.000335	0.000335	0.000335
0.00	1.000	0.000004	0.000004	0.000004

```
             .                           .

             .                           .

             .                           .

        Output for t = 4,...,16 removed

             .                           .

             .                           .

             .                           .
```

t	x	u1(x,t)	u2(x,t)	u3(x,t)
20.00	0.000	-0.000035	-0.000024	0.000041
20.00	0.100	-0.000218	-0.000321	0.000454
20.00	0.200	0.007326	0.001532	-0.005541
20.00	0.300	0.144030	0.091331	-0.162128
20.00	0.400	0.723723	0.560826	-0.877270
20.00	0.500	1.224314	0.998136	-1.508758
20.00	0.600	0.723723	0.560826	-0.877270
20.00	0.700	0.144030	0.091331	-0.162128
20.00	0.800	0.007326	0.001532	-0.005541
20.00	0.900	-0.000218	-0.000321	0.000454
20.00	1.000	-0.000035	-0.000024	0.000041

```
ncall =   363
```

Table 1.2: Abbreviated numerical output for `ncase=4`

We can note the following details about this output.

- As with Table 1.1, the values $t = 0, 4, \ldots, 20$ follow from the coding in the main program of Listing 1.1.

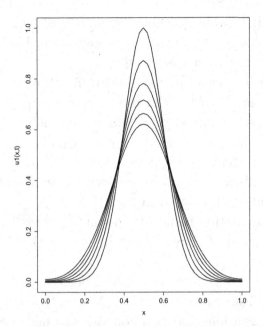

Figure 1.1a: Numerical solution of eq. (1.5a) for $u_1(x,t)$, ncase=1, from `matplot`

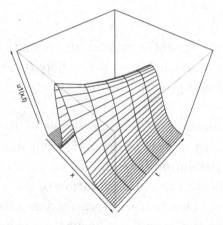

Figure 1.1b: Numerical solution of eq. (1.5a) for $u_1(x,t)$, ncase=1, from `persp`

- As with Table 1.1, the values $x = 0, 0.02, \ldots, 1$ follow from the coding in the main program of Listing 1.1, with every fifth value displayed.
- The solutions for u_1, u_2, u_3 are different as might expected for ncase=1 since the nine elements $a_{11}, \ldots, a_{33}$ taken from [2] have nonzero, asymmetrical values (with respect to the diagonal elements a_{11}, a_{22}, a_{33}).
- Rather than usual monotonic decay (dispersion) from just diffusion, the solutions with reaction and diffusion oscillate as indicated in Figs. 1.2.
- The oscillation results from complex eigenvalues (conjugate pairs) of the ODE/MOL system that are discussed in Chapter 2.
- The solutions oscillate between positive and negative values (Figs. 2.2). Since $u_1(x,t), u_2(x,t), u_3(x,t)$ represent concentrations (of morphogens), the negative values are accommodated (explained) by considering the solutions as oscillations (departures from, perturbations around) equilibrium (steady state) values ([2], p47).

The complex (figuratively and mathematically) solutions of Figs. 1.2 give an indication of the departure from just diffusion that is possible with the RD system of PDEs (Turing models).

Figs. 1.1 indicate a monotonic decay of the solution in t from diffusion (dispersion). Also, n=51 and nout=6 appear to give acceptable resolution in x and t.

Figs. 1.2 indicate the complex nature of the solutions of eqs. (1.5) for ncase=4. This can be interpreted as a form of dynamic pattern formation that would not occur with just diffusion. That is, a combination of reaction and diffusion is required, with selected values of $a_{11}, \ldots, a_{33}$ for reaction, and D_1, D_2, D_3 for diffusion.

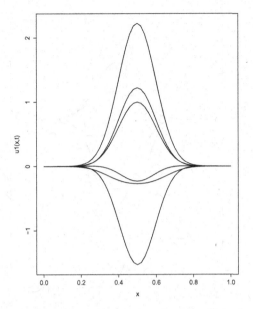

Figure 1.2a: Numerical solution of eq. (1.5a) for $u_1(x,t)$, ncase=4, from `matplot`

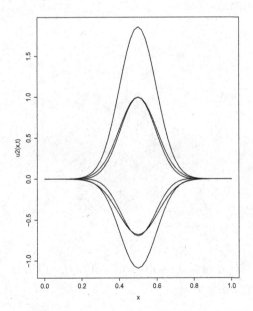

Figure 1.2b: Numerical solution of eq. (1.5b) for $u_2(x,t)$, ncase=4, from `matplot`

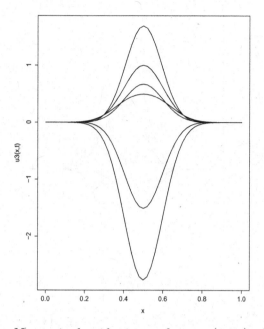

Figure 1.2c: Numerical solution of eq. (1.5c) for $u_3(x,t)$, ncase=4, from `matplot`

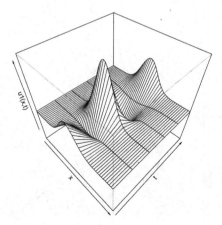

Figure 1.2d: Numerical solution of eq. (1.5a) for $u_1(x,t)$, ncase=4, from `persp`

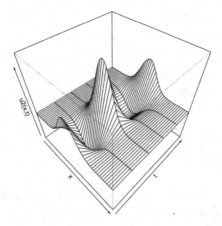

Figure 1.2e: Numerical solution of eq. (1.5b) for $u_2(x,t)$, ncase=4, from `persp`

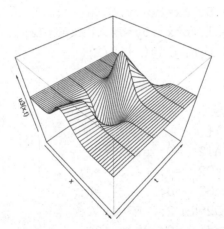

Figure 1.2f: Numerical solution of eq. (1.5c) for $u_3(x,t)$, ncase=4, from `persp`

To further demonstrate the oscillatory solutions to eqs. (1.5) for ncase=4, Listing 1.1 was modified slightly to plot the midpoint values $u_1(x = x_u/2, t), u_2(x = x_u/2, t), u_3(x = x_u/2, t)$ vs t. The modified listing follows.

```
#
# 1D 3 x 3 Turing
#
# Delete previous workspaces
  rm(list=ls(all=TRUE))
#
# Access ODE integrator
  library("deSolve");
#
# Access functions for numerical solution
  setwd("f:/turing/1D");
  source("pde_1b.R");
#
# Grid in x
  xl=0;xu=1;n=51;dx=(xu-xl)/(n-1);
  x=seq(from=xl,to=xu,by=dx);dxs=dx^2;
  n2=2*n;n3=3*n;n_mid=(n+1)/2;
#
# Parameters
  ncase=4;
  if(ncase==1){
    a11  =0; a12  =0; a13  =0;
    a21  =0; a22  =0; a23  =0;
    a31  =0; a32  =0; a33  =0;
    D1=dx^2; D2=dx^2;  D3=dx^2;
    t0   =0; tf   =2; nout=101;
    c1   =0; c2  =50;
    zl   =0; zu   =1;}
  if(ncase==2){
    a11= -1; a12  =0;  a13  =0;
    a21 = 0; a22 =-1;  a23 = 0;
    a31 = 0; a32  =0;  a33 =-1;
    D1=dx^2; D2=dx^2;  D3=dx^2;
    t0   =0; tf   =2; nout=101;
```

```
  c1    =0; c2   =50;
  zl    =0; zu   =1;}
if(ncase==3){
  a11   =1; a12   =0;   a13   =0;
  a21   =0; a22   =1;   a23   =0;
  a31   =0; a32   =0;   a33   =1;
  D1=dx^2; D2=dx^2;  D3=dx^2;
  t0    =0; tf    =2; nout=101;
  c1    =0; c2   =50;
  zl    =0; zu    =1;}
if(ncase==4){
  a11   =-10/3; a12       =3; a13    =-1;
  a21       =-2; a22     =7/3; a23      =0;
  a31       =3; a32      =-4; a33      =0;
  D1=2/3*dx^2; D2=1/3*dx^2; D3=0*dx^2;
  t0        =0; tf          =20; nout =101;
  c1        =0; c2         =50;
  zl        =-2; zu          =3;}
if(ncase==5){
  a11   =-1; a12    =-1; a13   = 0;
  a21   =1; a22     =0; a23    =-1;
  a31   =0; a32     =1; a33     =0;
  D1=1*dx^2; D2=0*dx^2; D3=0*dx^2;
  t0    =0; tf     =2; nout =101;
  c1    =0; c2    =50;
  zl    =-1; zu     =2;}
#
# Factors used in pde_1d.R
  D1dx2=D1/dx^2;D2dx2=D2/dx^2;D3dx2=D3/dx^2;
#
# Independent variable for ODE integration
  tout=seq(from=t0,to=tf,by=(tf-t0)/(nout-1));
#
# ICs
```

```
  u0=rep(0,n3);
  for(i in 1:n){
    u0[i]   =c1+exp(-c2*(x[i]-0.5)^2);
    u0[i+n] =c1+exp(-c2*(x[i]-0.5)^2);
    u0[i+n2]=c1+exp(-c2*(x[i]-0.5)^2);
  }
  ncall=0;
#
# ODE integration
  out=lsodes(y=u0,times=tout,func=pde_1b,
      sparsetype="sparseint",rtol=1e-6,
      atol=1e-6,maxord=5);
  nrow(out)
  ncol(out)
  cat(sprintf("\n n_mid = %2d\n",n_mid));
#
# Arrays for numerical solution
  u1=rep(0,nout);
  u2=rep(0,nout);
  u3=rep(0,nout);
  t=rep(0,nout);
  for(it in 1:nout){
    u1[it]=out[it,n_mid+1];
    u2[it]=out[it,n_mid+1+n];
    u3[it]=out[it,n_mid+1+n2];
       t[it]=out[it,1];
  }
  cat(sprintf(" ncall = %4d\n",ncall));
#
# Plot 2D numerical solution
  matplot(x,u1,type="l",lwd=2,col="black",
    lty=1,xlab="x",ylab="u1(x,t)",main="");
  matplot(x,u2,type="l",lwd=2,col="black",
    lty=1,xlab="x",ylab="u2(x,t)",main="");
```

```
matplot(x,u3,type="l",lwd=2,col="black",
  lty=1,xlab="x",ylab="u3(x,t)",main="");
```

Listing 1.3: Main program for eqs. (1.5) with plotting of
$$u_1(x = x_u/2, t), u_2(x = x_u/2, t), u_3(x = x_u/2, t)$$

The following differences between Listings 1.1 and 1.3 can be noted.

- The ODE/MOL routine is named pde_1b to give a second set of routines, but it is the same as pde_1a.
- A midpoint index n_mid is defined corresponding to $x = x_u/2$.

```
#
# Grid in x
  xl=0;xu=1;n=51;dx=(xu-xl)/(n-1);
  x=seq(from=xl,to=xu,by=dx);dxs=dx^2;
  n2=2*n;n3=3*n;n_mid=(n+1)/2;
```

- ncase=4 is used.

```
#
# Parameters
  ncase=4;
```

- The ODE integration is with pde_1b and n_mid is displayed.

```
#
# ODE integration
  out=lsodes(y=u0,times=tout,func=pde_1b,
      sparsetype="sparseint",rtol=1e-6,
      atol=1e-6,maxord=5);
  nrow(out)
  ncol(out)
  cat(sprintf("\n n_mid = %2d\n",n_mid));
```

- The midpoint values $u_1(x = x_u/2, t)$, $u_2(x = x_u/2, t)$, $u_3(x = x_u/2, t)$ are stored as a function of t for subsequent plotting.

```
#
# Arrays for numerical solution
  u1=rep(0,nout);
  u2=rep(0,nout);
  u3=rep(0,nout);
  t=rep(0,nout);
  for(it in 1:nout){
    u1[it]=out[it,n_mid+1];
    u2[it]=out[it,n_mid+1+n];
    u3[it]=out[it,n_mid+1+n2];
      t[it]=out[it,1];
  }
  cat(sprintf(" ncall = %4d\n",ncall));
```

- The labeling of the ordinate (vertical) axis reflects the midpoint values.

```
#
# Plot 2D numerical solution
  matplot(x,u1,type="l",lwd=2,col="black",
    lty=1,xlab="x",ylab="u1(x,t)",main="");
  matplot(x,u2,type="l",lwd=2,col="black",
    lty=1,xlab="x",ylab="u2(x,t)",main="");
  matplot(x,u3,type="l",lwd=2,col="black",
    lty=1,xlab="x",ylab="u3(x,t)",main="");
```

The numerical output follows, and the graphical output is in Figs. 3.1.

```
[1] 101

[1] 154

n_mid = 26

ncall = 353
```

Table 1.3: Abbreviated numerical output for ncase=4, midpoint plotting

The oscillation for ncase=4 is clear in Figs. 1.3, and contrasts with the monotone decay that would result from just diffusion (e.g., ncase=1). An eigenvalue analysis is given in Chapter 2 that explains the oscillation.

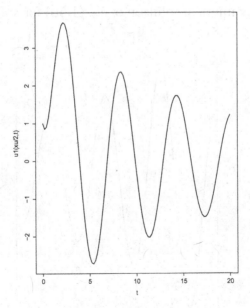

Figure 1.3a: Numerical solution of eq. (1.5a) for $u_1(x = x_u/2, t)$, ncase=4

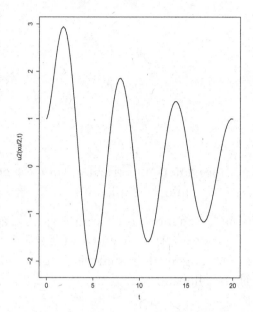

Figure 1.3b: Numerical solution of eq. (1.5b) for $u_2(x = x_u/2, t)$, ncase=4

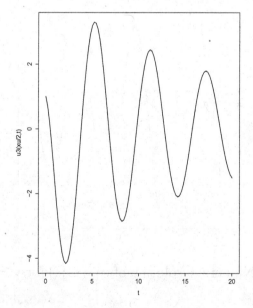

Figure 1.3c: Numerical solution of eq. (1.5c) for $u_3(x = x_u/2, t)$, ncase=4

References

[1] Schiesser, W.E. (2013), *Partial Differential Equation Analysis in Biomedical Engineering*, Cambridge University Press, Cambridge, UK

[2] Turing, A.M. (1952), The chemical basis of morphogenesis, *Philosophical Transactions of the Royal Society of London, Series B, Biological Sciences*, **237**, no. 641, 37-72

Chapter 2

Eigenvalue Analysis

If the reaction-diffusion (RD) PDEs discussed in Chapter 1 are linear, the method of lines (MOL) approximation of the PDEs gives a linear ODE system that can be analyzed by the basic methods of linear algebra. Emphasis is given in the following discussion to the approximate calculation of the ODE eigenvalues that provide an indication of the stability, time scale and form of the temporal (transient) response of the approximating MOL/ODEs.

(2.1) 1D 3 × 3 Linear Model

As an application of the MOL to eqs. (1.5), Turing [1] approximated the spatial derivatives with second-order finite differences (FDs).

$$\frac{du_{1,i}}{dt} = a_{11}u_{1,i} + a_{12}u_{2,i} + a_{13}u_{3,i} + D_1\left(u_{1,i+1} - 2u_{1,i} + u_{1,i-1}\right)$$

$$\text{(2.1a)}$$

$$\frac{du_{2,i}}{dt} = a_{21}u_{1,i} + a_{22}u_{2,i} + a_{23}u_{3,i} + D_2\left(u_{2,i+1} - 2u_{2,i} + u_{2,i-1}\right)$$

$$\text{(2.1b)}$$

$$\frac{du_{3,i}}{dt} = a_{31}u_{1,i} + a_{32}u_{2,i} + a_{33}u_{3,i} + D_3\left(u_{3,i+1} - 2u_{3,i} + u_{3,i-1}\right)$$

$$\text{(2.1c)}$$

$i = 1, 2, \ldots, n$ where n is the number of points in the spatial FD grid in x. Note that the factor $\dfrac{1}{\Delta x^2}$ of the FD approximations has been included in the diffusivities D_1, D_2, D_3 (according to Turing). This formulation of the FDs is explained for Listings 1.1 and 1.2 in Chapter 1.

Eqs. (2.1) are a system of linear, constant coefficient ODEs with the solution

$$u_{1,i}(t) = c_{1,i}e^{\lambda t} \tag{2.2a}$$

$$u_{2,i}(t) = c_{2,i}e^{\lambda t} \tag{2.2b}$$

$$u_{3,i}(t) = c_{3,i}e^{\lambda t} \tag{2.2c}$$

Substitution of eqs. (2.2) in eqs. (2.1), followed by cancellation[1] of the common factor $e^{\lambda t}$, gives a linear algebraic system

$$c_{1,i}\lambda = a_{11}c_{1,i} + a_{12}c_{2,i} + a_{13}c_{3,i} + D_1(c_{1,i+1} - 2c_{1,i} + c_{1,i-1}) \tag{2.3a}$$

$$c_{2,i}\lambda = a_{21}c_{1,i} + a_{22}c_{2,i} + a_{23}c_{3,i} + D_2(c_{2,i+1} - 2c_{2,i} + c_{2,i-1}) \tag{2.3b}$$

$$c_{3,i}\lambda = a_{31}c_{1,i} + a_{32}c_{2,i} + a_{33}c_{3,i} + D_3(c_{3,i+1} - 2c_{3,i} + c_{3,i-1}) \tag{2.3c}$$

Eqs. (2.3) are a $3n \times 3n$ system of homogeneous linear algebraic equations

$$(\mathbf{A} - \lambda\mathbf{I})\,\mathbf{c} = \mathbf{0} \tag{2.4a}$$

where

$\mathbf{A}$ $3n \times 3n$ coefficient matrix of eqs. (2.3)

$\mathbf{c}$ $3n -$ vector of unknowns of eqs. (2.3)

---

[1] The cancellation is possible since $e^{\lambda t}$ is nonzero for finite t.

I $3n \times 3n$ identity matrix

0 $3n$ − vector of zeros

λ eigenvalue in eqs. (2.2)

Eq. (2.4a) will have a nontrivial (**c** nonzero) solution if and only if the determinant of the coefficient matrix is zero.

$$|\mathbf{A} - \lambda\mathbf{I}| = 0 \qquad\qquad (2.4b)$$

Eq. (2.4b), the *characteristic equation* for eq. (2.4a), is a $3n$-order polynomial. Factoring of the polynomial gives the $3n$-vector of eigenvalues, $\lambda_i, i = 1, 2, \ldots, 3n$.

An R implementation of these basic operations in linear algebra is given in the next section.

(2.2) Eigenvalue Analysis of Linear Model

An examination of the eigenvalue vector from eq. (2.4b) gives an indication for the MOL/ODE system of the:

- Stability[2] (if the solution is unbounded for large t).
- Stiffness (the separation or range of values of the eigenvalue real parts)[3].
- Time scale of the solution, generally determined by the eigenvalue with the smallest $|Re\lambda|$ so that $\min(|Re\lambda|)t < 10$.

---

[2]If the real part of any of the eigenvalues is positive, the ODE system will be unstable. This follows from the analytical solution which is a linear combination of exponential functions of the form $e^{(Re\lambda + iIm\lambda)t}$ where λ is a particular eigenvalue and $i = \sqrt{-1}$.

[3]The ODE stiffness is often characterized by the *stiffness ratio*

$$SR = \frac{\max(|Re\lambda|)}{\min(|Re\lambda|)}.$$ An example of the SR is given subsequently.

Since the characteristic eq. (2.4b) is a $3n$-order polynomial with n the number of grid points in the MOL spatial grid, analytical factoring of the polynomial is not feasible and we resort to a numerical method which is based on a numerical approximation of the ODE system *Jacobian matrix*.

If the MOL/ODE system is

$$\frac{du_1}{dt} = f_1(u_1, u_2, \ldots, u_n)$$

$$\frac{du_2}{dt} = f_2(u_1, u_2, \ldots, u_n) \tag{2.5a}$$

$$\vdots$$

$$\frac{du_n}{dt} = f_n(u_1, u_2, \ldots, u_n)$$

the RHS function vector $(f_1, f_2, \ldots, f_n)^T$ is the basis for the Jacobian matrix

$$\mathbf{J} = \begin{bmatrix} \dfrac{\partial f_1}{\partial u_1} & \dfrac{\partial f_1}{\partial u_2} & \cdots & \dfrac{\partial f_1}{\partial u_n} \\[2ex] \dfrac{\partial f_2}{\partial u_1} & \dfrac{\partial f_2}{\partial u_2} & \cdots & \dfrac{\partial f_2}{\partial u_n} \\[2ex] & & \ddots & \\[1ex] \dfrac{\partial f_n}{\partial u_1} & \dfrac{\partial f_n}{\partial u_2} & \cdots & \dfrac{\partial f_n}{\partial u_n} \end{bmatrix} \tag{2.5b}$$

where n is the number of points in the MOL spatial grid.

The $n \times n$ partial derivatives of $\mathbf{J}$ can be approximated by finite differences (FDs).

$$\frac{\partial f_i}{\partial u_j} \approx$$

$$\frac{f_i(u_1, u_2, \ldots, u_j + \Delta u_j, \ldots, u_n) - f_i(u_1, u_2, \ldots, u_j, \ldots, u_n)}{\Delta u_j}$$

$$\tag{2.5c}$$

$i = 1, 2, \ldots, n$; $j = 1, 2, \ldots, n$ and Δu_j is a small, finite change (FD) in u_j. For a linear ODE system,

$$\frac{\partial f_i}{\partial u_j} = a_{ij} \tag{2.5d}$$

where a_{ij} is the ith row, jth column element of $\mathbf{A}$ of eq. (2.4b). That is, the Jacobian matrix of a linear ODE system is a constant matrix and therefore has to be computed only once (typically during the numerical integration of the ODE system). For a nonlinear ODE system, the elements of the Jacobian matrix are a function of the dependent variables $(u_1, u_2, \ldots, u_j, \ldots, u_n)^T$ (and therefore must be computed as the ODE solution evolves through t).

Eq. (2.5c) is the basis for the calculation of the numerical Jacobian matrix for the $3n \times 3n$ MOL/ODE system discussed in Chapter 1. We now consider some specific examples of the calculation of the numerical Jacobian matrix.

(2.3) Example 1: 2 × 2 ODE System

To start, a 2×2 linear constant coefficient ODE system is considered to illustrate some basic ideas in ODEs and linear algebra.

$$\frac{du_1}{dt} = -au_1 + bu_2; \quad u_1(t = 0) = 2 \tag{2.6a}$$

$$\frac{du_2}{dt} = bu_1 - au_2; \quad u_2(t = 0) = 0 \tag{2.6b}$$

where a, b are constants defined numerically in the main program of Listing 2.1.

Since eqs. (2.6) are linear, constant coefficient ODEs, they have solutions of the form

$$u_1(t) = c_1 e^{\lambda t}; \quad u_2(t) = c_2 e^{\lambda t} \tag{2.7a,b}$$

Substitution of eqs. (2.7) in eqs. (2.6) gives

$$\lambda c_1 e^{\lambda t} = -a c_1 e^{\lambda t} + b c_2 e^{\lambda t} \qquad (2.8a)$$

$$\lambda c_2 e^{\lambda t} = b c_1 e^{\lambda t} - a c_2 e^{\lambda t} \qquad (2.8b)$$

Division of eqs. (2.8) by $e^{\lambda t}$ and minor rearrangement gives

$$(-a - \lambda)c_1 + b c_2 = 0 \qquad (2.9a)$$

$$b c_1 + (-a - \lambda)c_2 = 0 \qquad (2.9b)$$

or

$$(\mathbf{A} - \lambda \mathbf{I})\mathbf{c} = \mathbf{0} \qquad (2.9c)$$

with

$$\mathbf{A} = \begin{bmatrix} a_{11} & a_{12} \\ a_{21} & a_{22} \end{bmatrix} = \begin{bmatrix} -a & b \\ b & -a \end{bmatrix} ; \ \mathbf{I} = \begin{bmatrix} 1 & 0 \\ 0 & 1 \end{bmatrix} ; \ \mathbf{c} = \begin{bmatrix} c_1 \\ c_2 \end{bmatrix} ; \ \mathbf{0} = \begin{bmatrix} 0 \\ 0 \end{bmatrix}$$

Eqs. (2.9) are linear, homogeneous algebraic equations for c_1, c_2 with a nontrival solution (c_1, c_2 nonzero) if and only if the determinant of the coefficient matrix is zero.

$$det \begin{bmatrix} a_{11} - \lambda & a_{12} \\ a_{21} & a_{22} - \lambda \end{bmatrix} = det \begin{bmatrix} -a - \lambda & b \\ b & -a - \lambda \end{bmatrix} = 0 \quad (2.10a)$$

or

$$(-a - \lambda)^2 - b^2 = 0$$

$$\lambda^2 + 2a\lambda + a^2 - b^2 = 0 \qquad (2.10b)$$

Eq. (2.10b) is the characteristic equation for eqs. (2.8) with the solution[4]

$$\lambda_1, \lambda_2 = \frac{-2a \pm \sqrt{(2a)^2 - 4(1)(a^2 - b^2)}}{2(1)} = -a \pm b \qquad (2.10c)$$

That is, $\lambda_1 = -a + b, \lambda_2 = -a - b$.

Eqs. (2.8) therefore have the solution

$$u_1(t) = c_{11}e^{-\lambda_1 t} + c_{12}e^{-\lambda_2 t} \qquad (2.11a)$$

$$u_2(t) = c_{21}e^{-\lambda_1 t} + c_{22}e^{-\lambda_2 t} \qquad (2.11b)$$

where the constant vectors

$$\mathbf{c_1} = \begin{bmatrix} c_{11} \\ c_{21} \end{bmatrix}_{\lambda_1} ; \ \mathbf{c_2} = \begin{bmatrix} c_{12} \\ c_{22} \end{bmatrix}_{\lambda_2}$$

are the eigenvectors for λ_1, λ_2.

$c_{11}, c_{12}, c_{21}, c_{22}$ are evaluated from either eq. (2.9a) or eq. (2.9b), and the ICs of eqs. (2.6). The final result is[5]

$$u_1(t) = e^{\lambda_1 t} + e^{\lambda_2 t} \qquad (2.12a)$$

$$u_2(t) = e^{\lambda_1 t} - e^{\lambda_2 t} \qquad (2.12b)$$

---

[4]Since **A** is symmetric, the eigenvalues are real. The choice $a_{11} = a_{22} = -a$, $a_{12} = a_{21} = b$ was made to simplify the calculation of λ_1, λ_2.

[5]Eqs. (12.2) can be verified by substitution in eqs. (2.6).

$$\lambda_1 e^{\lambda_1 t} + \lambda_2 e^{\lambda_2 t} = -a\left(e^{\lambda_1 t} + e^{\lambda_2 t}\right) + b\left(e^{\lambda_1 t} - e^{\lambda_2 t}\right)$$

$$\lambda_1 e^{\lambda_1 t} - \lambda_2 e^{\lambda_2 t} = b\left(e^{\lambda_1 t} + e^{\lambda_2 t}\right) - a\left(e^{\lambda_1 t} - e^{\lambda_2 t}\right)$$

or

$$(-a - \lambda_1 + b)\, e^{\lambda_1 t} = 0; \ (-a - \lambda_2 - b)\, e^{\lambda_2 t} = 0$$

$$(\lambda_1 - \lambda_1)\, e^{\lambda_1 t} = 0; \ (\lambda_2 - \lambda_2)\, e^{\lambda_2 t} = 0$$

Eqs. (2.12) also satisfy the ICs of eqs. (2.6).

that is

$$c_1 = \begin{bmatrix} 1 \\ 1 \end{bmatrix}_{\lambda_1} ; \; c_2 = \begin{bmatrix} 1 \\ -1 \end{bmatrix}_{\lambda_2}$$

(2.3.1) Main program

A main program for the implementation of eq. (2.5c) is listed next.

```
#
# 2 x 2 ODE
#
# Delete previous workspaces
  rm(list=ls(all=TRUE))
#
# Access functions for numerical solution
  setwd("f:/turing/jacob");
  source("eigen1.R");
#
# ODE parameters
  a=5.5;b=4.5;n=2;
#
# Base dependent variable vector
  ub=rep(1,n);
  for(i in 1:n){
    cat(sprintf("\n i = %2d  ub[i] = %8.4f",
                i,ub[i]));
  }
#
# Base dependent variable derivative vector
  utb=eigen1(t,ub,parm);
  for(i in 1:n){
    cat(sprintf("\n i = %2d  utb[i] = %8.4f",
                i,utb[i]));
  }
```

```
#
# Increment dependent variable vector
  u=rep(1,n);ut=rep(0,n);
  J=matrix(0,nrow=n,ncol=n);
#
# Step through columns
  for(j in 1:n){
#   u[j]=1.1*ub[j];
    u[j]=ub[j]+0.01;
    ut=eigen1(t,u,parm);
#
#   Step through rows
    for(i in 1:n){
      J[i,j]=(ut[i]-utb[i])/(u[j]-ub[j]);
      cat(sprintf("\n i = %2d   j = %2d
        J(i,j) = %8.4f",i,j,J[i,j]));
#
#   Next row
    }
    u[j]=ub[j];
    cat(sprintf("\n"));
#
# Next column
  }
#
# Compute and display eigenvalues
  lam=eigen(J,only.values=TRUE);
  lamVec = lam$values;
  for(i in 1:n){
    cat(sprintf(
      "\n i = %2d  Re = %8.4f  Im = %8.4f",
      i,Re(lamVec[i]),Im(lamVec[i])));
  }
```

Listing 2.1: Main program for eq. (2.5c)

We can note the following details about Listing 2.1.

- Previous workspaces are deleted. The `setwd` (set working directory) requires editing for the local computer to specify the directory (folder) with the R routines (note the use of / rather than the usual \). `eigen1` is a function for a 2×2 linear, constant coefficient ODE system (discussed next).

```
#
# 2 x 2 ODE
#
# Delete previous workspaces
  rm(list=ls(all=TRUE))
#
# Access functions for numerical solution
  setwd("f:/turing/jacob");
  source("eigen1.R");
```

- Constants a, b in the ODE system are defined numerically (and are passed to `eigen1` without special designation, a feature of R).

```
#
# ODE parameters
  a=5.5;b=4.5;n=2;
```

The eigenvalues are therefore from eq. (2.10c) $\lambda_1 = -a + b = -5.5 + 4.5 = -1$, $\lambda_2 = -a - b = -5.5 - 4.5 = -10$, which is a stable ODE system since the eigenvalues are real and negative.

- Base values `ub` for u_1, u_2 (eq. (2.5a)) are given unit values with the `rep` utility.

```
#
# Base dependent variable vector
```

```
ub=rep(1,n);
for(i in 1:n){
  cat(sprintf("\n i = %2d  ub[i] = %8.4f",
              i,ub[i]));
}
```

`cat(sprintf())` is a general purpose output statement for numerical values (`ub[i]`) with prescribed formats (`%2d` for integers, `%8.4f` for floating point numbers with a decimal point)[6].

- The derivative vector `utb` at the base values `ub` is evaluated by a call to `eigen1`.

```
#
# Base dependent variable derivative vector
utb=eigen1(t,ub,parm);
for(i in 1:n){
  cat(sprintf("\n i = %2d  utb[i] = %8.4f",
              i,utb[i]));
}
```

`utb[i]` corresponds to $f_i(u_1, u_2, \ldots, u_j, \ldots, u_n)$; $i = 1, 2, \ldots, n$ in eq. (2.5c).

- Vectors for f_i, u_j in eq. (2.5a) are declared (allocated).

```
#
# Increment dependent variable vector
u=rep(1,n);ut=rep(0,n);
J=matrix(0,nrow=n,ncol=n);
```

---

[6]`%2d` designates an integer with two figures (columns). `%8.4f` designates a floating point number with 8 columns, counting spaces, and 4 figures after the decimal point. Note also that the elements of vectors and arrays are defined with [] (`ub[i]`) rather than () which are used for the arguments of functions.

A matrix for the $n \times n$ Jacobian matrix $\mathbf{J}$ in eq. (2.5b) is declared with the `matrix` utility.

- $u_j + \Delta u_j$ in eq. (2.5c) is defined by a change in u_j. In the first case, a relative increment of 10% is used (`u[j]=1.1*ub[j]`). The relative increment does not work if `ub[j]=0`, so that in the second case, an absolute increment 0.01 is used (`u[j]=ub[j]+0.01`)[7].

```
#
# Step through columns
   for(j in 1:n){
#    u[j]=1.1*ub[j];
     u[j]=ub[j]+0.01;
     ut=eigen1(t,u,parm);
```

j is the index for columns in the Jacobian matrix and corresponds to j in eq. (2.5c). `eigen1` is used to calculate the corresponding derivative vector `ut`.

- Within a given column (j), the rows are varied to implement eq. (2.5c).

```
#
#    Step through rows
     for(i in 1:n){
        J[i,j]=(ut[i]-utb[i])/(u[j]-ub[j]);
        cat(sprintf("\n i = %2d   j = %2d
           J(i,j) = %8.4f",i,j,J[i,j]));
#
#    Next row
     }
     u[j]=ub[j];
```

---

[7]The two increments could be combined into a single statement `u[j]=1.1*ub[j]+0.01` which would handle `ub[j]` zero or nonzero.

```
    cat(sprintf("\n"));
#
# Next column
 }
```

i is the index for the rows of the Jacobian matrix and corresponds to i in eq. (2.5c). The computation of the i,jth element of the Jacobian matrix is

```
J[i,j]=(ut[i]-utb[i])/(u[j]-ub[j])
```

(from eq. (2.5c)).

- At this point, all $n \times n$ elements of the approximate Jacobian matrix have been computed. The n eigenvalues of **J** can now be computed (numerically, without factoring the characteristic polynomial (2.4b)).

```
#
# Compute and display eigenvalues
  lam=eigen(J,only.values=TRUE);
  lamVec = lam$values;
  for(i in 1:n){
    cat(sprintf(
       "\n i = %2d  Re = %8.4f   Im = %8.4f",
       i,Re(lamVec[i]),Im(lamVec[i])));
  }
```

The R routine **eigen** is used to compute the eigenvalues of **J**. In the call, only eigenvalues are specified (only.values=TRUE) rather than eigenvalues and eigenvectors (which is a time-consuming operation). **eigen** returns the eigenvalues as a **list** that is converted to numerical values with **$values**[8]. The real and imaginary

---

[8]The assistance of G. W. Griffiths with this conversion is gratefully acknowledged.

parts of the eigenvalues are displayed with the `Re` and `Im` utilities.

Function `eigen1` called above is now considered.

(2.3.2) Subordinate routine

```
  eigen1=function(t,u,parm){
#
# Function eigen1 computes the
# t derivative vector for u(t)
#
# Derivative vector
  ut=rep(0,n);
  ut[1]=-a*u[1]+b*u[2];
  ut[2]= b*u[1]-a*u[2];
#
# Return derivative vector
  return(c(ut));
  }
```

Listing 2.2: Routine <code>eigen1</code> for eqs. (2.6)

We can note the following details about Listing 2.2.

- The function is defined. `t` is the current value of t in eqs. (2.6). `u` is the 2-vector ODE of dependent variables. `parm` is an argument to pass parameters to `eigen1` (unused). The arguments must be listed in the order stated to properly interface with the main program of Listing 2.1. The composite derivative vector of the LHSs of eqs. (2.6) is calculated next and returned to the main program.

```
    eigen1=function(t,u,parm){
#
# Function eigen1 computes the
# t derivative vector for u(t)
```

- The derivative vector for eqs. (2.6) is calculated.

```
#
# Derivative vector
  ut=rep(0,n);
  ut[1]=-a*u[1]+b*u[2];
  ut[2]= b*u[1]-a*u[2];
```

- The derivative vector is returned to the main program. c is the R utility for a numerical vector.

```
#
# Return derivative vector
  return(c(ut));
  }
```

The final } concludes eigen1.

The numerical output from the routines in Listings 2.1 and 2.2 is next.

(2.3.3) Model output

Numerical output from execution of the main program and subordinate routine eigen1 follows.

```
i =  1  ub[i]  =    1.0000
i =  2  ub[i]  =    1.0000


i =  1  utb[i] =   -1.0000
i =  2  utb[i] =   -1.0000


i =  1  j =  1  J(i,j) =   -5.5000
i =  2  j =  1  J(i,j) =    4.5000


i =  1  j =  2  J(i,j) =    4.5000
i =  2  j =  2  J(i,j) =   -5.5000
```

```
i =  1  Re =  -1.0000  Im =    0.0000
i =  2  Re = -10.0000  Im =    0.0000
```

Table 2.1: Numerical output for eqs. (2.6)

We can note the following details about this output.

- The base values of u_1, u_2 are confirmed (from ub=rep(1,n)).

```
i =  1  ub[i] =    1.0000
i =  2  ub[i] =    1.0000
```

- The base values of the t derivatives of u_1, u_2 follow from eqs. (2.6).

```
i =  1  utb[i] =   -1.0000
i =  2  utb[i] =   -1.0000
```

From eqs. (2.6),

utb[1]=-5.5*1 + 4.5*1 = -1

utb[2]= 4.5*1 - 5.5*1 = -1

- The elements of the approximate Jacobian matrix are correct.

```
i =  1  j =  1  J(i,j) =   -5.5000
i =  2  j =  1  J(i,j) =    4.5000

i =  1  j =  2  J(i,j) =    4.5000
i =  2  j =  2  J(i,j) =   -5.5000
```

- The eigenvalues from function `eigen` are correct.

```
i =  1  Re =  -1.0000  Im =   0.0000
i =  2  Re = -10.0000  Im =   0.0000
```

From eq. (2.10c), $\lambda_1 = -a + b = -5.5 + 4.5 = -1$, $\lambda_2 = -a - b = -5.5 - 4.5 = -10$.

We can conclude from this output that the calculation of the approximate Jacobian matrix is coded correctly and therefore the calculation of the eigenvalues of larger ODE/MOL systems can be performed without having to factor the characteristic equation (polynomial). This application to higher order ODE/MOL systems is considered next. In particular, an eigenvalue analysis of the $3n \times 3n$ ODE/MOL system discussed in Chapter 1 is possible.

(2.4) Example 2: $3n \times 3n$ Reaction-diffusion System

An eigenvalue analysis of the 3 PDE reaction-diffusion (RD) system for $u_1(x,t), u_2(x,t), u_3(x,t)$ of eqs. (1.5) discussed in Chapter 1 is now considered. Since this system is linear, the Jacobian matrix is constant. If any of the $3n \times 3n$ eigenvalues have positive real parts, the PDE system is unstable. Of particular interest is the stability of the Turing systems for `ncase=4,5` in the main program of Listing 1.1.

(2.4.1) Main program

A main program for the eigenvalue analysis of eqs. (1.5) for `ncase=1,2,3,4,5` follows.

```
#
# 1D 3 x 3 Turing
#
# Delete previous workspaces
  rm(list=ls(all=TRUE))
```

```
#
# Access functions for numerical solution
  setwd("f:/turing/jacob");
  source("eigen2.R");
#
# Grid in x
  xl=0;xu=1;n=11;dx=(xu-xl)/(n-1);
  x=seq(from=xl,to=xu,by=dx);dxs=dx^2;
#
# Parameters
  ncase=1;
  if(ncase==1){
    a11= 0; a12= 0;  a13=0;
    a21= 0; a22= 0;  a23=0;
    a31= 0; a32= 0;  a33=0;
    D1=dxs; D2=dxs; D3=dxs;}
  if(ncase==2){
    a11=-1; a12= 0; a13= 0;
    a21= 0; a22=-1; a23= 0;
    a31= 0; a32= 0; a33=-1;
    D1=dxs; D2=dxs; D3=dxs;}
  if(ncase==3){
    a11= 1; a12= 0; a13= 0;
    a21= 0; a22= 1; a23= 0;
    a31= 0; a32= 0; a33= 1;
    D1=dxs; D2=dxs; D3=dxs;}
  if(ncase==4){
  a11=-10/3 ; a12=3       ; a13=-1  ;
  a21=-2     ; a22=7/3    ; a23=0   ;
  a31=3      ; a32=-4     ; a33=0   ;
  D1=2/3*dxs; D2=1/3*dxs; D3=0*dxs;}
  if(ncase==5){
  a11=-1  ; a12=-1  ; a13= 0  ;
  a21= 1  ; a22= 0  ; a23=-1  ;
```

```
   a31= 0   ; a32= 1   ; a33= 0   ;
   D1=1*dxs;  D2=0*dxs;  D3=0*dxs;}
#
# Base dependent variable vector
   n3=3*n;
   ub=rep(0,n3);
# for(i in 1:n3){
#   cat(sprintf("\n i = %2d  ub[i] = %8.4f",
#               i,ub[i]));
# }
#
# Base dependent variable derivative vector
   utb=eigen2(t,ub,parm);
# for(i in 1:n3){
#   cat(sprintf("\n i = %2d  utb[i] = %8.4f",
#               i,utb[i]));
# }
#
# Incremented dependent variable vector
   u=rep(0,n3);ut=rep(0,n3);
   J=matrix(0,nrow=n3,ncol=n3);
#
# Step through Jacobian matrix columns
   for(j in 1:n3){
     u[j]=ub[j]+0.01;
     ut=eigen2(t,u,parm);
#
#    Step through Jacobian matrix rows
     for(i in 1:n3){
       J[i,j]=(ut[i]-utb[i])/(u[j]-ub[j]);
#      cat(sprintf(
#        "\n i = %2d  j = %2d  J(i,j) = %8.4f",
#        i,j,J[i,j]));
#
```

```
#    Next row
     }
     u[j]=ub[j];
#    cat(sprintf("\n"));
#
# Next column
     }
#
# Compute and display eigenvalues
     lam=eigen(J,only.values=TRUE);
     lamVec = lam$values;
     Re_lam=rep(0,n3);Im_lam=rep(0,n3);
     iout=0;
     for(i in 1:n3){
       Re_lam[i]=Re(lamVec[i]);
       Im_lam[i]=Im(lamVec[i]);
       if(Re_lam[i]>0){
         cat(sprintf(
           "\n i = %2d   Re = %8.4f   Im = %8.4f",
           i,Re_lam[i],Im_lam[i]));
         iout=iout+1;
       }
     }
     cat(sprintf("\n iout = %3d",iout));
#
# Plot eigenvalues
     Im_max=max(Im_lam);Im_min=-Im_max;
     Re_0=rep(0,n3);
     plot(Re_lam,Im_lam,lwd=2,col="black",pch="o",
         ylim=c(Im_min,Im_max),xlab="Re_lam",
         ylab="Im_lam",main="");
     lines(Re_0,Im_lam,type="l",lwd=2);
```

Listing 2.3: Main program for eqs. (1.5)

We can note the following details about Listing 2.3.

- Previous workspaces are deleted. The setwd (set working directory) requires editing for the local computer to specify the directory (folder) with the R routines (note the use of / rather than the usual \). eigen2 is a function for the $3n \times 3n$ linear, constant coefficient MOL system for eqs. (1.5) (discussed next).

```
#
# 1D 3 x 3 Turing
#
# Delete previous workspaces
  rm(list=ls(all=TRUE))
#
# Access functions for numerical solution
  setwd("f:/turing/jacob");
  source("eigen2.R");
```

- A uniform grid in x of 11 points is defined with the seq utility for the interval $x_l = 0 \leq x \leq x_u = 1$. Therefore, the vector x has the values $x = 0, 1/10, \ldots, 1$.

```
#
# Grid in x
  xl=0;xu=1;n=11;dx=(xu-xl)/(n-1);
  x=seq(from=xl,to=xu,by=dx);dxs=dx^2;
```

The small number of grid points, $n = 11$, was selected to keep the output (if selected by deactivating comments) at a more manageable level. This main program also executes with $n = 51$ (used in Chapter 1) as discussed subsequently.

- Five cases are programmed as in Listing 1.1.

```
#
# Parameters
  ncase=1;
  if(ncase==1){
     a11= 0; a12= 0;   a13=0;
     a21= 0; a22= 0;   a23=0;
     a31= 0; a32= 0;   a33=0;
     D1=dxs; D2=dxs; D3=dxs;}
  if(ncase==2){
     a11=-1; a12= 0; a13= 0;
     a21= 0; a22=-1; a23= 0;
     a31= 0; a32= 0; a33=-1;
     D1=dxs; D2=dxs; D3=dxs;}
  if(ncase==3){
     a11= 1; a12= 0; a13= 0;
     a21= 0; a22= 1; a23= 0;
     a31= 0; a32= 0; a33= 1;
     D1=dxs; D2=dxs; D3=dxs;}
  if(ncase==4){
  a11=-10/3 ; a12=3      ; a13=-1  ;
  a21=-2     ; a22=7/3   ; a23=0   ;
  a31=3      ; a32=-4    ; a33=0   ;
  D1=2/3*dxs; D2=1/3*dxs; D3=0*dxs;}
  if(ncase==5){
  a11=-1  ; a12=-1  ; a13= 0  ;
  a21= 1  ; a22= 0  ; a23=-1  ;
  a31= 0  ; a32= 1  ; a33= 0  ;
  D1=1*dxs; D2=0*dxs; D3=0*dxs;}
```

- Base values ub for u_1, u_2, u_3 (eqs. (1.5)) are given zero values with the **rep** utility.

```
#
# Base dependent variable vector
  n3=3*n;
  ub=rep(0,n3);
# for(i in 1:n3){
#   cat(sprintf("\n i = %2d  ub[i] = %8.4f",
#                i,ub[i]));
# }
```

The output is suppressed because of limited space in the discussion.

- The derivative vector utb at the base values ub is evaluated by a call to eigen2.

```
#
# Base dependent variable derivative vector
  utb=eigen2(t,ub,parm);
# for(i in 1:n3){
#   cat(sprintf("\n i = %2d  utb[i] = %8.4f",
#                i,utb[i]));
# }
```

The output is suppressed because of limited space in the discussion.

- Vectors for f_i, u_j in eq. (2.5a) are declared (allocated).

```
#
# Incremented dependent variable vector
  u=rep(0,n3);ut=rep(0,n3);
  J=matrix(0,nrow=n3,ncol=n3);
```

A matrix for the $n \times n$ Jacobian matrix $\mathbf{J}$ in eq. (2.5b) is declared with the matrix utility.

- $u_j + \Delta u_j$ in eq. (2.5c) is defined by a change in u_j. An absolute increment 0.01 is used (u[j]=ub[j]+0.01)

```
#
# Step through Jacobian matrix columns
  for(j in 1:n3){
    u[j]=ub[j]+0.01;
    ut=eigen2(t,u,parm);
```

j is the index for columns in the Jacobian matrix and corresponds to j in eq. (2.5c). eigen2 is used to calculate the corresponding derivative vector ut.

- Within a given column (j), the rows are varied to implement eq. (2.5c). i is the index for the rows of the Jacobian matrix and corresponds to i in eq. (2.5c). The computation of the i,jth element of the Jacobian matrix is

```
#
#    Step through Jacobian matrix rows
     for(i in 1:n3){
       J[i,j]=(ut[i]-utb[i])/(u[j]-ub[j]);
#      cat(sprintf(
#        "\n i = %2d  j = %2d  J(i,j) = %8.4f",
#        i,j,J[i,j]));
#
#    Next row
     }
     u[j]=ub[j];
#    cat(sprintf("\n"));
#
# Next column
   }
```

The output is suppressed because of limited space in the discussion.

- At this point, all $3n \times 3n$ elements of the approximate Jacobian matrix have been computed. The $3n$ eigenvalues of **J** can now be computed (numerically, without factoring the characteristic polynomial (2.4b)).

```
#
# Compute and display eigenvalues
  lam=eigen(J,only.values=TRUE);
  lamVec = lam$values;
  Re_lam=rep(0,n3);Im_lam=rep(0,n3);
  iout=0;
  for(i in 1:n3){
    Re_lam[i]=Re(lamVec[i]);
    Im_lam[i]=Im(lamVec[i]);
    if(Re_lam[i]>0){
      cat(sprintf(
        "\n i = %2d  Re = %8.4f  Im = %8.4f",
        i,Re_lam[i],Im_lam[i]));
      iout=iout+1;
    }
  }
    cat(sprintf("\n iout = %3d",iout));
```

A test is performed for an unstable eigenvalue with a positive real part (`if(Re_lam[i]>0)`). If one is found, the counter for unstable eigenvalues, `iout`, is incremented, then displayed after all of the eigenvalues are tested for instability.

- The eigenvalues are plotted as points (circles, `pch="o"`) in the complex plane. A line is added (`lines`) for a zero real part (the boundary between stable and unstable eigenvalues) using two vectors, `Re_0`, `Im_0`.

```
#
# Plot eigenvalues
  Re_0=rep(0,n3);Im_0=rep(0,n3);
  Im_0[1]=-1;Im_0[n3]=1;
```

```
plot(Re_lam,Im_lam,lwd=2,col="black",pch="o",
    ylim=c(Im_min,Im_max),xlab="Re_lam",
    ylab="Im_lam",main="");
lines(Re_0,Im_0,type="l",lwd=2);
```

Automatic scaling of the abscissa (horizontal, x) and ordinate (vertical, y) axes accommodates variations in the plots for ncase=1,...,5 (Figs. 2.1 to 2.5).

The subordinate routine eigen2 called by the main program of Listing 2.3 is considered next.

(2.4.2) Subordinate routine

The subordinate routine for the ODEs (2.1) follows.

```
eigen2=function(t,u,parm){
#
# Function eigen2 computes the t derivative
# vector for u1(x,t), u2(x,t), u3(x,t)
#
# One vector to three vectors
  u1 =rep(0,n);u2 =rep(0,n);u3 =rep(0,n);
  u1t=rep(0,n);u2t=rep(0,n);u3t=rep(0,n);
  for(i in 1:n){
    u1[i]=u[i];
    u2[i]=u[i+n];
    u3[i]=u[i+2*n];
  }
#
# u1t(x,t)
  for(i in 1:n){
    if(i==1){u1t[1]=a11*u1[1]+a12*u2[1]+a13*u3[1]+
           2*D1*(u1[  2]-u1[1])/dx^2;}
    if(i==n){u1t[n]=a11*u1[n]+a12*u2[n]+a13*u3[n]+
           2*D1*(u1[n-1]-u1[n])/dx^2;}
    if((i>1)&(i<n)){
```

```
      u1t[i]=a11*u1[i]+a12*u2[i]+a13*u3[i]+
             D1*(u1[i+1]-2*u1[i]+u1[i-1])/dx^2;}
  }
#
# u2t(x,t)
  for(i in 1:n){
    if(i==1){u2t[1]=a21*u1[1]+a22*u2[1]+a23*u3[1]+
             2*D2*(u2[   2]-u2[1])/dx^2;}
    if(i==n){u2t[n]=a21*u1[n]+a22*u2[n]+a23*u3[n]+
             2*D2*(u2[n-1]-u2[n])/dx^2;}
    if((i>1)&(i<n)){
      u2t[i]=a21*u1[i]+a22*u2[i]+a23*u3[i]+
             D2*(u2[i+1]-2*u2[i]+u2[i-1])/dx^2;}
  }
#
# u3t(x,t)
  for(i in 1:n){
    if(i==1){u3t[1]=a31*u1[1]+a32*u2[1]+a33*u3[1]+
             2*D3*(u3[   2]-u3[1])/dx^2;}
    if(i==n){u3t[n]=a31*u1[n]+a32*u2[n]+a33*u3[n]+
             2*D3*(u3[n-1]-u3[n])/dx^2;}
    if((i>1)&(i<n)){
      u3t[i]=a31*u1[i]+a32*u2[i]+a33*u3[i]+
             D3*(u3[i+1]-2*u3[i]+u3[i-1])/dx^2;}
  }
#
# Three vectors to one vector
  ut=rep(0,n3);
  for(i in 1:n){
    ut[i]     =u1t[i];
    ut[i+n]   =u2t[i];
    ut[i+2*n]=u3t[i];
  }
#
```

```
# Return derivative vector
  return(c(ut));
  }
```

Listing 2.4: Routine `eigen2` for eqs. (2.6)

`eigen2` is very similar to `pde_1a` of Listing 1.2. Only the differences are noted next.

- The function is defined. `t` is the current value of t in eqs. (2.1). `u` is the $3n = 3(11) = 33$-vector ODE dependent variables ($n = 51$ is used in `pde_1a` of Listing 1.2 rather than $n = 11$ in Listing 2.4).

```
  eigen2=function(t,u,parm){
#
# Function eigen2 computes the t derivative
# vector for u1(x,t), u2(x,t), u3(x,t)
```

- The derivative vector is returned to the main program. `c` is the R utility for a numerical vector (a `list` is also used in `pde_1a` of Listing 1.2 as required by `lsodes`).

```
#
# Return derivative vector
  return(c(ut));
  }
```

The final `}` concludes `eigen2`.

The numerical output from the routines in Listings 2.3 and 2.4 is next.

(2.4.3) Model output

For `ncase=1` in Listing 2.3, the output is

```
i = 31   Re =    0.0000   Im =    0.0000
i = 32   Re =    0.0000   Im =    0.0000
```

```
i = 33  Re =   0.0000  Im =   0.0000
```

```
iout =   3
```

Table 2.1: Output for `ncase=1`, Listing 2.3

These three eigenvalues are displayed as a consequence of the statements

```
if(Re_lam[i]>0){
  cat(sprintf(
    "\n i = %2d  Re = %8.4f  Im = %8.4f",
    i,Re_lam[i],Im_lam[i]));
  iout=iout+1;
}
```

The real part of eigenvalues 31,32,33 are zero, but with a small positive value in floating point arithmetic, possibly resulting from the finite difference (FD) approximation of the three spatial derivatives $\frac{\partial^2 u_1}{\partial x^2}, \frac{\partial^2 u_2}{\partial x^2}, \frac{\partial^2 u_3}{\partial x^2}$. However, the system of $3n \times 3n$ ODEs for `ncase=1` is stable, as expected for diffusion only. The plot of the eigenvalues in Fig. 2.1 also indicates that all of the 33 eigenvalues have nonpositive real parts and zero imaginary parts. The monotonic decay for diffusion only is reflected in Figs. 1.1a,b.

The vertical line from

```
lines(Re_0,Im_0,type="l",lwd=2);
```

assists in determining if all of the eigenvalues have $Re\lambda \le 0$ (if all of the eigenvalues are in the left half of the complex plane).

For `ncase=2` in Listing 2.3, the output is

```
iout =   0
```

Table 2.2: Output for `ncase=2`, Listing 2.3

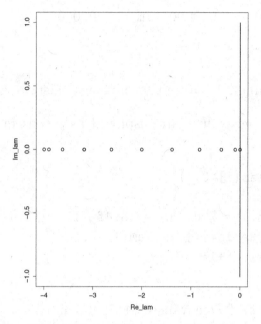

Figure 2.1: Eigenvalues in the complex plane, `ncase=1`

No eigenvalues are displayed since the real parts are clearly all negative, as reflected in Fig. 2.2. This is expected since the reaction terms from

```
if(ncase==2){
  a11=-1; a12= 0; a13= 0;
  a21= 0; a22=-1; a23= 0;
  a31= 0; a32= 0; a33=-1;
```

reflect a consumption by the chemical reactions that adds to the dispersion of the diffusion. The more negative real parts are indicated in Fig. 2.2 (note the values along the real axis).

This can be confirmed by displaying the full set of eigenvalues with

```
for(i in 1:n3){
  Re_lam[i]=Re(lamVec[i]);Im_lam[i]=Im(lamVec[i]);
#   if(Re_lam[i]>0){
```

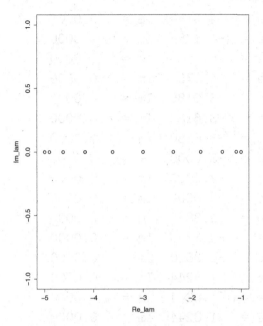

Figure 2.2: Eigenvalues in the complex plane, `ncase=2`

```
cat(sprintf(
   "\n i = %2d   Re = %8.4f   Im = %8.4f",
   i,Re_lam[i],Im_lam[i]));
#     iout=iout+1;
#   }
 }
```

so that the output is

```
i =  1  Re =  -5.0000  Im =     0.0000
i =  2  Re =  -5.0000  Im =     0.0000
i =  3  Re =  -5.0000  Im =     0.0000
i =  4  Re =  -4.9021  Im =     0.0000
i =  5  Re =  -4.9021  Im =     0.0000
i =  6  Re =  -4.9021  Im =     0.0000
i =  7  Re =  -4.6180  Im =     0.0000
i =  8  Re =  -4.6180  Im =     0.0000
```

```
i =  9   Re =   -4.6180   Im =    0.0000
i = 10   Re =   -4.1756   Im =    0.0000
i = 11   Re =   -4.1756   Im =    0.0000
i = 12   Re =   -4.1756   Im =    0.0000
i = 13   Re =   -3.6180   Im =    0.0000
i = 14   Re =   -3.6180   Im =    0.0000
i = 15   Re =   -3.6180   Im =    0.0000
i = 16   Re =   -3.0000   Im =    0.0000
i = 17   Re =   -3.0000   Im =    0.0000
i = 18   Re =   -3.0000   Im =    0.0000
i = 19   Re =   -2.3820   Im =    0.0000
i = 20   Re =   -2.3820   Im =    0.0000
i = 21   Re =   -2.3820   Im =    0.0000
i = 22   Re =   -1.8244   Im =    0.0000
i = 23   Re =   -1.8244   Im =    0.0000
i = 24   Re =   -1.8244   Im =    0.0000
i = 25   Re =   -1.3820   Im =    0.0000
i = 26   Re =   -1.3820   Im =    0.0000
i = 27   Re =   -1.3820   Im =    0.0000
i = 28   Re =   -1.0979   Im =    0.0000
i = 29   Re =   -1.0979   Im =    0.0000
i = 30   Re =   -1.0979   Im =    0.0000
i = 31   Re =   -1.0000   Im =    0.0000
i = 32   Re =   -1.0000   Im =    0.0000
i = 33   Re =   -1.0000   Im =    0.0000
```

The stiffness ratio is

$$SR = \frac{|Re\lambda|_{max}}{|Re\lambda|_{min}} = \frac{5}{1} = 5$$

so the system is not stiff.

For `ncase=3` in Listing 2.3 the output is

```
i = 19   Re =    1.0000   Im =    0.0000
i = 20   Re =    1.0000   Im =    0.0000
i = 21   Re =    1.0000   Im =    0.0000
```

```
i = 22   Re =    0.9021   Im =    0.0000
i = 23   Re =    0.9021   Im =    0.0000
i = 24   Re =    0.9021   Im =    0.0000
i = 25   Re =    0.6180   Im =    0.0000
i = 26   Re =    0.6180   Im =    0.0000
i = 27   Re =    0.6180   Im =    0.0000
i = 31   Re =    0.1756   Im =    0.0000
i = 32   Re =    0.1756   Im =    0.0000
i = 33   Re =    0.1756   Im =    0.0000

iout =   12
```

Table 2.3: Output for **ncase=3**, Listing 2.3

The real part of eigenvalues **19** to **33** is positive so the system is unstable. This is demonstrated in Fig. 2.3. Note that each eigenvalue is repeated three times.

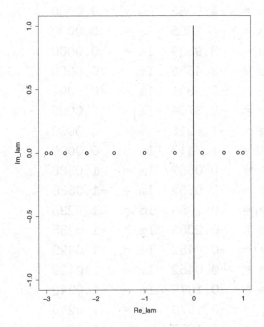

Figure 2.3: Eigenvalues in the complex plane, **ncase=3**

Again, the line from

```
lines(Re_0,Im_0,type="l",lwd=2);
```

identifies the eigenvalues in the right complex plane $Re\lambda > 0$. The instability results from the positive reaction terms in

```
if(ncase==3){
  a11= 1;  a12= 0;  a13= 0;
  a21= 0;  a22= 1;  a23= 0;
  a31= 0;  a32= 0;  a33= 1;
  D1=dxs;  D2=dxs;  D3=dxs;}
```

The reactions therefore produce more reactants with increasing t (u_1, u_2, u_3 increase with t).

For ncase=4 in Listing 2.3, the output with the display of the full set of eigenvalues is

```
i =  1   Re =   -4.5835   Im =    0.0000
i =  2   Re =   -4.5063   Im =    0.0000
i =  3   Re =   -4.2806   Im =    0.0000
i =  4   Re =   -3.9239   Im =    0.0000
i =  5   Re =   -3.4630   Im =    0.0000
i =  6   Re =   -2.9321   Im =    0.0000
i =  7   Re =   -2.3704   Im =    0.0000
i =  8   Re =   -1.8215   Im =    0.0000
i =  9   Re =   -1.3413   Im =    0.0000
i = 10   Re =   -0.0552   Im =    1.0588
i = 11   Re =   -0.0552   Im =   -1.0588
i = 12   Re =   -0.2083   Im =    1.0235
i = 13   Re =   -0.2083   Im =   -1.0235
i = 14   Re =   -0.0452   Im =    1.0429
i = 15   Re =   -0.0452   Im =   -1.0429
i = 16   Re =   -0.1979   Im =    1.0240
i = 17   Re =   -0.1979   Im =   -1.0240
i = 18   Re =   -0.1687   Im =    1.0249
```

```
i = 19    Re =   -0.1687    Im =   -1.0249
i = 20    Re =   -0.1259    Im =    1.0239
i = 21    Re =   -0.1259    Im =   -1.0239
i = 22    Re =   -0.0775    Im =    1.0192
i = 23    Re =   -0.0775    Im =   -1.0192
i = 24    Re =   -0.0203    Im =    1.0148
i = 25    Re =   -0.0203    Im =   -1.0148
i = 26    Re =   -0.0339    Im =    1.0109
i = 27    Re =   -0.0339    Im =   -1.0109
i = 28    Re =   -1.0075    Im =    0.0000
i = 29    Re =   -0.0058    Im =    1.0024
i = 30    Re =   -0.0058    Im =   -1.0024
i = 31    Re =   -0.0014    Im =    1.0008
i = 32    Re =   -0.0014    Im =   -1.0008
i = 33    Re =   -0.8896    Im =    0.0000
```

Table 2.4: Output for `ncase=4`, Listing 2.3

We can note the following details about this output.

- The real part of all of the eigenvalues is negative so the system is stable.

- The eigenvalues now have nonzero imaginary parts so the system is oscillatory[9]. This is in contrast to the effect of diffusion which is nonoscillatory, and is a distinguishing feature of the Turing model ([1], p53) with the specific constants[10]

---

[9]The term *Turing instability* is frequently used in the literature. However, the solution is not unstable (all of the eigenvalue real parts are nonpositive) so a more suitable term might be *Turing oscillation*.

[10]Since the ODE coefficient matrix is no longer symmetric (as it was in `ncase=1,2,3`), the ODE Jacobian matrix eigenvalues are no longer necessarily real (as in `ncase=1,2,3`).

```
if(ncase==4){
a11=-10/3 ; a12=3      ; a13=-1  ;
a21=-2    ; a22=7/3    ; a23=0   ;
a31=3     ; a32=-4     ; a33=0   ;
D1=2/3*dxs; D2=1/3*dxs; D3=0*dxs;}
```

- The complex eigenvalues appear as conjugate pairs which is a consequence of a real characteristic polynomial, e.g.,

```
i = 10   Re =   -0.0552   Im =    1.0588
i = 11   Re =   -0.0552   Im =   -1.0588
```

- The stiffness ratio is

$$SR = \frac{|Re\lambda|_{max}}{|Re\lambda|_{min}} = \frac{4.5835}{0.0014} = 3274$$

so the system is substantially stiffer than for `ncase=2`. However, the integrator `lsodes` used in Chapter 1 can accommodate stiff ODE systems.

The complex eigenvalues are clearly demonstrated in Fig. 2.4.

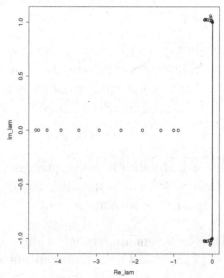

Figure 2.4: Eigenvalues in the complex plane, `ncase=4`

The line from

```
lines(Re_0,Im_lam,type="l",lwd=2);
```

also demonstrates that the eigenvalues are stable conjugate pairs.

The oscillatory solution for `ncase=4` is reflected in Figs. 1.2a,b,c,d,e,f and 1.3a,b,c.

For `ncase=5` in Listing 2.3, the output with the display of the full set of eigenvalues is

```
i =  1  Re =  -4.8003  Im =   0.0000
i =  2  Re =  -4.6985  Im =   0.0000
i =  3  Re =  -4.4020  Im =   0.0000
i =  4  Re =  -3.9370  Im =   0.0000
i =  5  Re =  -3.3435  Im =   0.0000
i =  6  Re =  -2.6717  Im =   0.0000
i =  7  Re =  -1.9795  Im =   0.0000
i =  8  Re =  -1.3457  Im =   0.0000
i =  9  Re =  -0.2151  Im =   1.3071
i = 10  Re =  -0.2151  Im =  -1.3071
i = 11  Re =  -0.2275  Im =   1.2868
i = 12  Re =  -0.2275  Im =  -1.2868
i = 13  Re =  -0.2482  Im =   1.2243
i = 14  Re =  -0.2482  Im =  -1.2243
i = 15  Re =  -0.2394  Im =   1.1395
i = 16  Re =  -0.2394  Im =  -1.1395
i = 17  Re =  -0.2012  Im =   1.0783
i = 18  Re =  -0.2012  Im =  -1.0783
i = 19  Re =  -0.1642  Im =   1.0469
i = 20  Re =  -0.1642  Im =  -1.0469
i = 21  Re =  -0.1373  Im =   1.0311
i = 22  Re =  -0.1373  Im =  -1.0311
i = 23  Re =  -0.1193  Im =   1.0229
i = 24  Re =  -0.1193  Im =  -1.0229
i = 25  Re =  -0.1080  Im =   1.0185
```

```
i = 26   Re =   -0.1080   Im =   -1.0185
i = 27   Re =   -0.1018   Im =    1.0164
i = 28   Re =   -0.1018   Im =   -1.0164
i = 29   Re =   -0.0998   Im =    1.0157
i = 30   Re =   -0.0998   Im =   -1.0157
i = 31   Re =   -0.8856   Im =    0.0000
i = 32   Re =   -0.6430   Im =    0.0000
i = 33   Re =   -0.5698   Im =    0.0000
```

Table 2.5: Output for `ncase=5`, Listing 2.3

We can note the following details about this output.

- The real part of all of the eigenvalues is negative so the system is stable.
- As in `ncase=4`, the eigenvalues for the specific constants ([1], p54) have nonzero imaginary parts so the system is oscillatory.

```
if(ncase==5){
a11=-1  ; a12=-1  ; a13= 0  ;
a21= 1  ; a22= 0  ; a23=-1  ;
a31= 0  ; a32= 1  ; a33= 0  ;
D1=1*dxs; D2=0*dxs; D3=0*dxs;}
```

- The complex eigenvalues appear as conjugate pairs which is a consequence of a real characteristic polynomial, e.g.,

```
i =  9   Re =   -0.2151   Im =    1.3071
i = 10   Re =   -0.2151   Im =   -1.3071
```

- The stiffness ratio is

$$SR = \frac{|Re\lambda|_{max}}{|Re\lambda|_{min}} = \frac{4.8003}{0.0998} = 48.1$$

so the system is not stiff. The wide variation in the stiffness ratio (3274 for `ncase=4`, 48.1 for `ncase=5`) indicates the value of a stiff ODE integrator (`lsodes`).

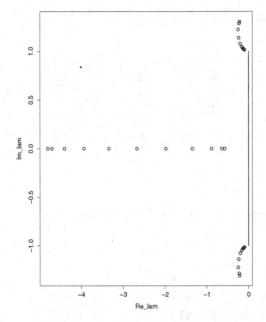

Figure 2.5: Eigenvalues in the complex plane, `ncase=5`

The complex eigenvalues are demonstrated in Fig. 2.5.
The line from

```
lines(Re_0,Im_0,type="l",lwd=2);
```

also demonstrates that the eigenvalues are stable conjugate pairs.

To conclude this discussion of the $3n \times 3n$ (three morphogen) system, we can consider an increase in the number of spatial grid points, e.g., $n = 11$ to $n = 51$. The only change in Listing 2.3 (with n=51 and display of all of the eigenvalues) is

```
#
# Grid in x
  xl=0;xu=1;n=51;dx=(xu-xl)/(n-1);
  x=seq(from=xl,to=xu,by=dx);dxs=dx^2;
```

The numerical output for **ncase=4** is

```
i =  1  Re =  -4.5835  Im =     0.0000
i =  2  Re =  -4.5804  Im =     0.0000
i =  3  Re =  -4.5711  Im =     0.0000
i =  4  Re =  -4.5556  Im =     0.0000
i =  5  Re =  -4.5339  Im =     0.0000
              .                  .
              .                  .
              .                  .

       Output for i=6 to 39 removed
              .                  .
              .                  .
              .                  .

i = 40  Re =  -1.4284  Im =     0.0000
i = 41  Re =  -1.3413  Im =     0.0000
i = 42  Re =  -1.2600  Im =     0.0000
i = 43  Re =  -1.1853  Im =     0.0000
i = 44  Re =  -1.1180  Im =     0.0000
i = 45  Re =  -0.0552  Im =     1.0588
i = 46  Re =  -0.0552  Im =    -1.0588
i = 47  Re =  -0.0548  Im =     1.0581
i = 48  Re =  -0.0548  Im =    -1.0581
i = 49  Re =  -1.0585  Im =     0.0000
              .                  .
              .                  .
              .                  .

     Output for i=50 to 140 removed
              .                  .
              .                  .
              .                  .

i = 141  Re =  -0.0014  Im =     1.0008
i = 142  Re =  -0.0014  Im =    -1.0008
i = 143  Re =  -0.0009  Im =     1.0004
```

```
i = 144   Re =   -0.0009   Im =   -1.0004
i = 145   Re =   -0.0002   Im =    1.0001
i = 146   Re =   -0.0002   Im =   -1.0001
i = 147   Re =   -0.0001   Im =    1.0000
i = 148   Re =   -0.0001   Im =   -1.0000
i = 149   Re =   -0.9653   Im =    0.0000
i = 150   Re =   -0.9323   Im =    0.0000
i = 151   Re =   -0.9086   Im =    0.0000
i = 152   Re =   -0.8944   Im =    0.0000
i = 153   Re =   -0.8896   Im =    0.0000
```

Table 2.6: Abbreviated output for `ncase=4`, $n = 51$, Listing 2.3

The $3(51) = 153$ eigenvalues are displayed in Fig. 2.6 which indicates that all of the real parts are nonpositive.

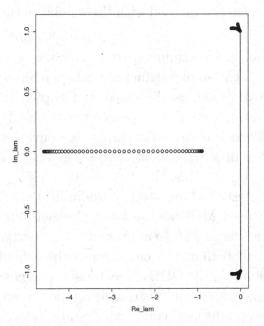

Figure 2.6: Eigenvalues in the complex plane, `ncase=4`, n=51

The stiffness ratio is

$$SR = \frac{|Re\lambda|_{max}}{|Re\lambda|_{min}} = \frac{4.5835}{0.0001} = 45835$$

so that the stiffness of the ODE system has increased substantially with $n = 51$ rather than $n = 11$. In general ODE/MOL systems are stiffer with increasing grid points (greater spatial resolution) which indicates the value of a stiff integrator for MOL analysis.

(2.5) Summary and Conclusions

The preceding examples of ODE eigenvalue analysis demonstrate the stability and oscillation of RD systems, and in particular, the Turing oscillation. Since the spatial derivatives $\frac{\partial^2 u_1}{\partial x^2}$, $\frac{\partial^2 u_2}{\partial x^2}$, $\frac{\partial^2 u_3}{\partial x^2}$ are approximated with three point, second order FDs (as suggested in [1]), an extension with higher order FDs would be instructive (to determine spatial accuracy or convergence). This is an application of p refinement where p refers to the order of the approximation, as discussed in Chapter 1. A variation in the number of spatial grid points would also be instructive (again, to determine spatial accuracy or convergence). This is an application of h refinement where h refers to the power of the spatial interval, $O(\Delta x^h)$.

These suggestions for further study indicate the flexibility of a computer-based MOL and associated eigenvalue analysis. In particular, the use of FDs to approximate the partial derivatives in the ODE Jacobian matrix circumvents the difficulty of formulating and factoring the ODE characteristic polynomial. In other words, the eigenvalue analysis can be applied to an ODE/MOL system of essentially any complexity, and can even be applied to nonlinear PDEs as discussed in the next chapter.

Reference

[1] Turing, A.M. (1952), The chemical basis of morphogenesis, *Philosophical Transactions of the Royal Society of London, Series B, Biological Sciences,* **237**, no. 641, 37-72

Chapter 3

Nonlinear Models

The method of lines (MOL) solution discussed in Chapter 1, and the eigenvalue analysis discussed in Chapter 2 pertain to a linear reaction-diffusion (RD) partial differential equation (PDE) model. A central result is the oscillation that is possible with a linear RD system, and which raises the question whether oscillation is also possible with a nonlinear RD system. Since a nonlinear system can be considered as a generalization of a linear system, as discussed in this chapter, we would expect that nonlinear oscillation would be possible, and this is confirmed subsequently.

(3.1) $1D$ 3 × 3 Nonlinear Model

The 3×3 linear model of eqs. (1.5) is modified to include nonlinear *chemotaxis* diffusion in place of the linear (Fickian) diffusion in eq. (3.1a)[1].

$$\frac{\partial u_1}{\partial t} = a_{11}u_1 + a_{12}u_2 + a_{13}u_3 + D_1\frac{\partial(u1\frac{\partial u_2}{\partial x})}{\partial x} - D_1\frac{\partial(u1\frac{\partial u_3}{\partial x})}{\partial x} \tag{3.1a}$$

---

[1]The linear diffusion term $D_1\frac{\partial^2 u_1}{\partial x^2}$ can easily be retained in eq. (3.1a).

$$\frac{\partial u_2}{\partial t} = a_{21}u_1 + a_{22}u_2 + a_{23}u_3 + D_2\frac{\partial^2 u_2}{\partial x^2} \qquad (3.1b)$$

$$\frac{\partial u_3}{\partial t} = a_{31}u_1 + a_{32}u_2 + a_{33}u_3 + D_3\frac{\partial^2 u_3}{\partial x^2} \qquad (3.1c)$$

The term $+D_1\dfrac{\partial(u1\frac{\partial u_2}{\partial x})}{\partial x}$ includes a *repellent* flux with the gra-

dient $\dfrac{\partial u_2}{\partial x}$ (as a consequence of the $+$). The term $-D_1\dfrac{\partial(u1\frac{\partial u_3}{\partial x})}{\partial x}$

includes an *attractant* flux with the gradient $\dfrac{\partial u_3}{\partial x}$ (as a conse-

quence of the $-$). Therefore, eq. (3.1a) is based on *chemotaxis attractant-repellent* diffusion[2].

The MOL solution of eqs. (3.1) is considered next.

(3.1.1) Main program

The main program for eqs. (3.1) is similar to the main program of Listing 1.1.

```
#
# 1D 3 x 3 RD
#
# Delete previous workspaces
  rm(list=ls(all=TRUE))
#
# Access ODE integrator
  library("deSolve");
#
# Access functions for numerical solution
```

---

[2]The chemotaxis terms are nonlinear as a consequence of the products between u_1 and the gradients $\dfrac{\partial u_2}{\partial x}, \dfrac{\partial u_3}{\partial x}$.

```
  setwd("f:/turing/1D");
  source("pde_1c.R");
#
# Grid in x
  xl=0;xu=1;n=51;dx=(xu-xl)/(n-1);
  x=seq(from=xl,to=xu,by=dx);
  n2=2*n;n3=3*n;
#
# Parameters
  ncase=4;
  if(ncase==4){
  a11   =-10/3; a12      =3; a13   =-1;
  a21      =-2; a22    =7/3; a23   =0;
  a31       =3; a32     =-4; a33   =0;
  D1=2/3*dx^2; D2=1/3*dx^2; D3=0*dx^2;
  t0        =0; tf       =20; nout  =6;
  c1        =0; c2       =50;
  zl       =-2; zu       =3;}
  if(ncase==5){
  a11    =-1; a12    =-1; a13   = 0;
  a21     =1; a22     =0; a23   =-1;
  a31     =0; a32     =1; a33    =0;
  D1=1*dx^2; D2=0*dx^2; D3=0*dx^2;
  t0     =0; tf    =20; nout   =6;
  c1     =0; c2    =50;
  zl    =-1; zu     =2;}
#
# Factors used in pde_1c.R
  D1dx2=D1/dx^2;D2dx2=D2/dx^2;D3dx2=D3/dx^2;
#
# Independent variable for ODE integration
  tout=seq(from=t0,to=tf,by=(tf-t0)/(nout-1));
#
# ICs
```

```
  u0=rep(0,n3);
  for(i in 1:n){
    u0[i]   =c1+exp(-c2*(x[i]-0.5)^2);
#   u0[i+n] =c1+exp(-c2*(x[i]-0.5)^2);
#   u0[i+n2]=c1+exp(-c2*(x[i]-0.5)^2);
    u0[i+n] =0;
    u0[i+n2]=0;
  }
  ncall=0;
#
# ODE integration
  out=lsodes(y=u0,times=tout,func=pde_1c,
      sparsetype="sparseint",rtol=1e-6,
      atol=1e-6,maxord=5);
  nrow(out)
  ncol(out)
#
# Arrays for numerical solution
  u1=matrix(0,nrow=n,ncol=nout);
  u2=matrix(0,nrow=n,ncol=nout);
  u3=matrix(0,nrow=n,ncol=nout);
  t=rep(0,nout);
  for(it in 1:nout){
  for(i  in 1:n){
    u1[i,it]=out[it,i+1];
    u2[i,it]=out[it,i+1+n];
    u3[i,it]=out[it,i+1+n2];
      t[it]=out[it,1];
  }
  }
#
# Display selected output
  for(it in 1:nout){
    cat(sprintf("\n      t          x    u1(x,t)
```

```
                    u2(x,t)    u3(x,t)\n"));
      iv=seq(from=1,to=n,by=5);
      for(i in iv){
       cat(sprintf(
         "%6.2f%9.3f%10.6f%10.6f%10.6f\n",
         t[it],x[i],u1[i,it],u2[i,it],u3[i,it]));
       }
       cat(sprintf("\n"));
      }
    cat(sprintf(" ncall = %4d\n",ncall));
#
# Plot 2D numerical solution
    matplot(x,u1,type="l",lwd=2,col="black",lty=1,
      xlab="x",ylab="u1(x,t)",main="");
    matplot(x,u2,type="l",lwd=2,col="black",lty=1,
      xlab="x",ylab="u2(x,t)",main="");
    matplot(x,u3,type="l",lwd=2,col="black",lty=1,
      xlab="x",ylab="u3(x,t)",main="");
#
# Plot 3D numerical solution
    persp(x,t,u1,theta=45,phi=45,xlim=c(xl,xu),
          ylim=c(t0,tf),xlab="x",ylab="t",
          zlab="u1(x,t)");
    persp(x,t,u2,theta=45,phi=45,xlim=c(xl,xu),
          ylim=c(t0,tf),xlab="x",ylab="t",
          zlab="u2(x,t)");
    persp(x,t,u3,theta=45,phi=45,xlim=c(xl,xu),
          ylim=c(t0,tf),xlab="x",ylab="t",
          zlab="u3(x,t)");
```

Listing 3.1: Main program for eqs. (3.1)

We can note the following details about Listing 3.1.

- Previous workspaces are removed. Then the ODE integrator library deSolve is accessed. The setwd (set working directory) uses / rather than the usual \, and requires editing on the local computer (to specify the directory with the R files).

```
#
# 1D 3 x 3 Turing
#
# Delete previous workspaces
  rm(list=ls(all=TRUE))
#
# Access ODE integrator
  library("deSolve");
#
# Access functions for numerical solution
  setwd("f:/turing/1D");
  source("pde_1c.R");
```

pde_1c is the routine for the MOL ODEs (discussed subsequently).

- A uniform grid in x of 51 points is defined with the seq utility for the interval $x_l = 0 \le x \le x_u = 1$. Therefore, the vector x has the values $x = 0, 1/50, \ldots, 1$.

```
#
# Grid in x
  xl=0;xu=1;n=51;dx=(xu-xl)/(n-1);
  x=seq(from=xl,to=xu,by=dx);
  n2=2*n;n3=3*n;
```

- Two cases are programmed with variations in the model parameters, i.e., ncase=4,5.

```
#
# Parameters
  ncase=4;
  if(ncase==4){
  a11  =-10/3; a12     =3; a13   =-1;
```

```
a21      =-2; a22     =7/3; a23      =0;
a31      =3; a32      =-4; a33       =0;
D1=2/3*dx^2; D2=1/3*dx^2; D3=0*dx^2;
t0       =0; tf        =20; nout     =6;
c1       =0; c2        =50;
zl       =-2; zu       =3;}
if(ncase==5){
a11      =-1; a12      =-1; a13      = 0;
a21      =1; a22       =0; a23       =-1;
a31      =0; a32       =1; a33       =0;
D1=1*dx^2; D2=0*dx^2; D3=0*dx^2;
t0       =0; tf        =20; nout     =6;
c1       =0; c2        =50;
zl       =-1; zu       =2;}
```

Cases `ncase=1,2,3` have been removed from Listing 1.1 in Chapter 1. To reiterate from Chapter 1,

- `ncase==4`: The reaction constants $a_{11}, \ldots, a_{33}$ and diffusivities are taken from Turing ([2], p53, eqs. (8.5)).
- `ncase==5`: The reaction constants $a_{11}, \ldots, a_{33}$ and diffusivities are taken from Turing ([2], p54, eqs. (8.7)).

A central idea by Turing is that the reaction coefficients $a_{11}, \ldots, a_{33}$ and diffusivities D_1, D_2, D_3 are selected so that some of the eigenvalues are complex (conjugate pairs) and the solution therefore oscillates in t. This is a marked distinction from Fickian diffusion which disperses the solution monotonically in t.

• Factors used in the ODE/MOL routine `pde_1c` are defined.

```
#
# Factors used in pde_1c.R
  D1dx2=D1/dx^2;D2dx2=D2/dx^2;D3dx2=D3/dx^2;
```

Each term has a division by dx^2 which cancels the same factor in the definition of D1,D2,D3 so that these diffusivities correspond to their use by Turing ([2], p47, eq. 6.2 applied to three components u_1, u_2, u_3).

- A uniform grid in t of 6 output points is defined with the seq utility for the interval $t_0 = 0 \leq t \leq t_f = 20$. Therefore, the vector tout has the values $t = 0, 4, \ldots, 20$.

```
#
# Independent variable for ODE integration
  tout=seq(from=t0,to=tf,by=(tf-t0)/(nout-1));
```

At this point, the intervals of x and t in eq. (3.1) are defined.

- IC functions for eqs. (3.1) are defined. n2,n3 are constants defined previously.

```
#
# ICs
  u0=rep(0,n3);
  for(i in 1:n){
    u0[i]    =c1+exp(-c2*(x[i]-0.5)^2);
#   u0[i+n]  =c1+exp(-c2*(x[i]-0.5)^2);
#   u0[i+n2] =c1+exp(-c2*(x[i]-0.5)^2);
    u0[i+n]  =0;
    u0[i+n2] =0;
  }
  ncall=0;
```

The functions for u_2, u_3 can also be activated (uncommented) to implement IC Gaussian functions. This not done in the first execution of the main program (Listing 3.1) since the gradient of u_3=u0[i+n2] in the

attractant chemotaxis term in eq. (3.1a) tends to destabilize the solution (explained subsequently). By starting
u3 at zero, this effect is minimized.

Another possibility would be to use randomly distributed IC values to investigate patterning in the solutions, but this might require an increase in the number
of grid points in x (above n=51) for improved spatial
resolution.

The counter for the calls to pde_1c is initialized (and
passed to pde_1c without a special designation).

- The system of $3(51) = 153$ MOL/ODEs is integrated
  by the library integrator lsodes (available in deSolve)
  with the sparse matrix option specified. As expected, the
  inputs to lsodes are the ODE function, pde_1c, the IC
  vector u0, and the vector of output values of t, tout.
  The length of u0 (e.g., 153) informs lsodes how many
  ODEs are to be integrated. func,y,times are reserved
  names.

```
#
# ODE integration
  out=lsodes(y=u0,times=tout,func=pde_1c,
      sparsetype="sparseint",rtol=1e-6,
      atol=1e-6,maxord=5);
  nrow(out)
  ncol(out)
```

The numerical solution to the ODEs is returned in matrix
out. In this case, out has the dimensions $nout \times (n+1) =$
6×154. The offset $n+1$ is required since the first element
of each column has the output t (also in tout), and the
$2, \ldots, n+1 = 2, \ldots, 154$ column elements have the 153
ODE solutions. This indexing of out in used next.

- The ODE solution is placed in 3 51×6 matrices, u1,u2,u3, for subsequent plotting (by stepping through the solution with respect to x and t within a pair of fors).

```
#
# Arrays for numerical solution
  u1=matrix(0,nrow=n,ncol=nout);
  u2=matrix(0,nrow=n,ncol=nout);
  u3=matrix(0,nrow=n,ncol=nout);
  t=rep(0,nout);
  for(it in 1:nout){
  for(i  in 1:n){
    u1[i,it]=out[it,i+1];
    u2[i,it]=out[it,i+1+n];
    u3[i,it]=out[it,i+1+n2];
      t[it]=out[it,1];
  }
  }
```

- The numerical solutions are displayed.

```
#
# Display selected output
  for(it in 1:nout){
    cat(sprintf("\n      t        x   u1(x,t)
              u2(x,t)   u3(x,t)\n"));
    iv=seq(from=1,to=n,by=5);
    for(i in iv){
    cat(sprintf(
      "%6.2f%9.3f%10.6f%10.6f%10.6f\n",
      t[it],x[i],u1[i,it],u2[i,it],u3[i,it]));
    }
    cat(sprintf("\n"));
  }
  cat(sprintf(" ncall = %4d\n",ncall));
```

To conserve space, only every fifth value in x of the solutions is displayed numerically (using the subscript iv).

- The solutions to eqs. (3.1), $u_1(x,t), u_2(x,t), u_3(x,t)$, are plotted vs x with t as a parameter in 2D with matplot.

```
#
# Plot 2D numerical solution
  matplot(x,u1,type="l",lwd=2,col="black",
    lty=1,xlab="x",ylab="u1(x,t)",main="");
  matplot(x,u2,type="l",lwd=2,col="black",
    lty=1,xlab="x",ylab="u2(x,t)",main="");
  matplot(x,u3,type="l",lwd=2,col="black",
    lty=1,xlab="x",ylab="u3(x,t)",main="");
```

Note that the rows of x (nrows=n=51) equals the rows of u1,u2,u3.

- The solutions to eqs. (3.1), $u_1(x,t), u_2(x,t), u_3(x,t)$, are plotted vs x and t in 3D perspective with persp.

```
#
# Plot 3D numerical solution
  persp(x,t,u1,theta=45,phi=45,xlim=c(xl,xu),
        ylim=c(t0,tf),xlab="x",ylab="t",
        zlab="u1(x,t)");
  persp(x,t,u2,theta=45,phi=45,xlim=c(xl,xu),
        ylim=c(t0,tf),xlab="x",ylab="t",
        zlab="u2(x,t)");
  persp(x,t,u3,theta=45,phi=45,xlim=c(xl,xu),
        ylim=c(t0,tf),xlab="x",ylab="t",
        zlab="u3(x,t)");
```

Automatic scaling in z is used. If the three plots are to have a common vertical scale in z, zlim=c(zl,zu) defined previously could be used. x,t have dimensions in agreement with u1,u2,u3 (nrows=n=51, ncols=nout=6).

The ODE/MOL routine called in the main program of Listing 3.1, pde_1c, follows.

(3.1.2) Subordinate routine

The ODE/MOL routine for eqs. (3.1) is listed next.

```
  pde_1c=function(t,u,parm){
#
# Function pde_1c computes the t derivative
# vector for u1(x,t), u2(x,t), u3(x,t)
#
# One vector to three vectors
  u1 =rep(0,n);u2 =rep(0,n);u3 =rep(0,n);
  u1t=rep(0,n);u2t=rep(0,n);u3t=rep(0,n);
  for(i in 1:n){
    u1[i]=u[i];
    u2[i]=u[i+n];
    u3[i]=u[i+n2];
  }
#
# u1t(x,t)
  for(i in 1:n){
    if(i==1){u1t[1]=a11*u1[1]+a12*u2[1]+a13*u3[1];}
    if(i==n){u1t[n]=a11*u1[n]+a12*u2[n]+a13*u3[n];}
    if((i>1)&(i<n)){
    u1t[i]=a11*u1[i]+a12*u2[i]+a13*u3[i]+
      D1dx2*((u1[i]+u1[i+1])/2*(u2[i+1]-u2[i])-
            (u1[i]+u1[i-1])/2*(u2[i]-u2[i-1]))-
      D1dx2*((u1[i]+u1[i+1])/2*(u3[i+1]-u3[i])-
            (u1[i]+u1[i-1])/2*(u3[i]-u3[i-1]));}
```

```
}
#
# u2t(x,t)
  for(i in 1:n){
    if(i==1){u2t[1]=a21*u1[1]+a22*u2[1]+a23*u3[1]+
            2*D2dx2*(u2[  2]-u2[1]);}
    if(i==n){u2t[n]=a21*u1[n]+a22*u2[n]+a23*u3[n]+
            2*D2dx2*(u2[n-1]-u2[n]);}
    if((i>1)&(i<n)){
      u2t[i]=a21*u1[i]+a22*u2[i]+a23*u3[i]+
            D2dx2*(u2[i+1]-2*u2[i]+u2[i-1]);}
  }
#
# u3t(x,t)
  for(i in 1:n){
    if(i==1){u3t[1]=a31*u1[1]+a32*u2[1]+a33*u3[1]+
            2*D3dx2*(u3[  2]-u3[1]);}
    if(i==n){u3t[n]=a31*u1[n]+a32*u2[n]+a33*u3[n]+
            2*D3dx2*(u3[n-1]-u3[n]);}
    if((i>1)&(i<n)){
      u3t[i]=a31*u1[i]+a32*u2[i]+a33*u3[i]+
            D3dx2*(u3[i+1]-2*u3[i]+u3[i-1]);}
  }
#
# Three vectors to one vector
  ut=rep(0,n3);
  for(i in 1:n){
    ut[i]   =u1t[i];
    ut[i+n] =u2t[i];
    ut[i+n2]=u3t[i];
  }
#
# Increment calls to pde_1c
  ncall <<- ncall+1;
```

```
#
# Return derivative vector
  return(list(c(ut)));
  }
```

Listing 3.2: ODE/MOL routine pde_1c for eqs. (3.1)

We can note the following details about pde_1c.

- The function is defined.

  ```
    pde_1c=function(t,u,parm){
  #
  # Function pde_1c computes the t derivative
  # vector for u1(x,t), u2(x,t), u3(x,t)
  ```

  t is the current value of t in eqs. (3.1). u is the 153-vector of ODE/MOL dependent variables. parm is an argument to pass parameters to pde_1c (unused, but required in the argument list). The arguments must be listed in the order stated to properly interface with lsodes called in the main program of Listing 3.1. The composite derivative vector of the LHSs of eqs. (3.1) is calculated next and returned to lsodes.

- The dependent variable vectors are placed in three vectors to facilitate the programming of eqs. (3.1).

  ```
  #
  # One vector to three vectors
    u1 =rep(0,n);u2 =rep(0,n);u3 =rep(0,n);
    u1t=rep(0,n);u2t=rep(0,n);u3t=rep(0,n);
    for(i in 1:n){
      u1[i]=u[i];
      u2[i]=u[i+n];
      u3[i]=u[i+n2];
    }
  ```

Vectors are also defined for the LHS derivatives in t of eqs. (3.1).

- $\dfrac{\partial u_1}{\partial t}$ of eq. (3.1a) is programmed in a for that steps through x, for(i in 1:n).

```
#
# u1t(x,t)
  for(i in 1:n){
    if(i==1){u1t[1]=a11*u1[1]+a12*u2[1]+
                    a13*u3[1];}
    if(i==n){u1t[n]=a11*u1[n]+a12*u2[n]+
                    a13*u3[n];}
    if((i>1)&(i<n)){
  u1t[i]=a11*u1[i]+a12*u2[i]+a13*u3[i]+
  D1dx2*((u1[i]+u1[i+1])/2*(u2[i+1]-u2[i])-
         (u1[i]+u1[i-1])/2*(u2[i]-u2[i-1]))-
  D1dx2*((u1[i]+u1[i+1])/2*(u3[i+1]-u3[i])-
         (u1[i]+u1[i-1])/2*(u3[i]-u3[i-1]));}
  }
```

This coding requires some additional explanation.

- At $x = x_l = 0$ $(i = 1)$, the homogeneous Neumann BCs

$$\frac{\partial u_2(x = 0, t)}{\partial x} = 0; \quad \frac{\partial u_3(x = 0, t)}{\partial x} = 0$$

are used. For the chemotaxis term in eq. (3.1a),

$$+D_1 \frac{\partial \left(u1(x = 0, t) \dfrac{\partial u_2(x = 0, t)}{\partial x} \right)}{\partial x}$$

$$-D_1 \frac{\partial \left(u1(x = 0, t) \dfrac{\partial u_3(x = 0, t)}{\partial x} \right)}{\partial x}$$

$$= 0$$

and the chemotaxis term at $x = x_l = 0$ $(i = 1)$ is zero. This is reflected in the coding

```
if(i==1){u1t[1]=a11*u1[1]+a12*u2[1]+
                a13*u3[1];}
```

(only the reaction terms remain).

– The same reasoning applies to the application of homogeneous Neumann BCs at the right boundary $x = x_u = 1$ $(i = n)$.

$$\frac{\partial u_2(x = 1, t)}{\partial x} = 0; \quad \frac{\partial u_3(x = 1, t)}{\partial x} = 0$$

and the chemotaxis term at $x = x_u = 1$ $(i = n)$ is zero. This is reflected in the coding

```
if(i==n){u1t[n]=a11*u1[n]+a12*u2[n]+
                a13*u3[n];}
```

(again, only the reaction terms remain)[2].

– For the interior points $x_l + \Delta x \le x \le x_u - \Delta x$, the reaction terms are retained. The chemotaxis terms in eq. (3.1a),

$$+D_1\frac{\partial(u_1\frac{\partial u_2}{\partial x})}{\partial x} - D_1\frac{\partial(u_1\frac{\partial u_3}{\partial x})}{\partial x}$$

are approximated with FDs.

$$+D_1\frac{\partial(u_1\frac{\partial u_2}{\partial x})}{\partial x} \approx$$

$$D_1\left[\left(\frac{u_1(i+1) + u_1(i)}{2}\right)\left(\frac{u_2(i+1) - u_2(i)}{\Delta x}\right)\right.$$

---

[2]As an alternative, FD approximations could be used at the boundaries $x = x_l, x_u$.

$$-\left(\frac{u_1(i)+u_1(i-1)}{2}\right)\left(\frac{u_2(i)-u_2(i-1)}{\Delta x}\right)\Bigg]$$
$$\overline{\hspace{3cm}\Delta x\hspace{3cm}}$$

$$=\frac{D_1}{\Delta x^2}\left[\left(\frac{u_1(i+1)+u_1(i)}{2}\right)(u_2(i+1)-u_2(i))\right.$$

$$\left.-\left(\frac{u_1(i)+u_1(i-1)}{2}\right)(u_2(i)-u_2(i-1))\right]$$

$$-D_1\frac{\partial(u_1\dfrac{\partial u_3}{\partial x})}{\partial x}\approx$$

$$-D_1\left[\left(\frac{u_1(i+1)+u_1(i)}{2}\right)\left(\frac{u_3(i+1)-u_3(i)}{\Delta x}\right)\right.$$

$$\left.-\left(\frac{u_1(i)+u_1(i-1)}{2}\right)\left(\frac{u_3(i)-u_3(i-1)}{\Delta x}\right)\right]$$
$$\overline{\hspace{3cm}\Delta x\hspace{3cm}}$$

$$=\frac{D_1}{\Delta x^2}\left[\left(\frac{u_1(i+1)+u_1(i)}{2}\right)(u_3(i+1)-u_3(i))\right.$$

$$\left.-\left(\frac{u_1(i)+u_1(i-1)}{2}\right)(u_3(i)-u_3(i-1))\right]$$

The corresponding coding is

```
    if((i>1)&(i<n)){
u1t[i]=a11*u1[i]+a12*u2[i]+a13*u3[i]+
D1dx2*((u1[i]+u1[i+1])/2*(u2[i+1]-u2[i])-
    (u1[i]+u1[i-1])/2*(u2[i]-u2[i-1]))-
D1dx2*((u1[i]+u1[i+1])/2*(u3[i+1]-u3[i])-
    (u1[i]+u1[i-1])/2*(u3[i]-u3[i-1]));}
```

- The coding for $u_2(x,t)$ (with homogeneous Neumann BCs) is

```
#
# u2t(x,t)
  for(i in 1:n){
    if(i==1){u2t[1]=a21*u1[1]+a22*u2[1]+
      a23*u3[1]+2*D2dx2*(u2[  2]-u2[1]);}
    if(i==n){u2t[n]=a21*u1[n]+a22*u2[n]+
      a23*u3[n]+2*D2dx2*(u2[n-1]-u2[n]);}
    if((i>1)&(i<n)){
      u2t[i]=a21*u1[i]+a22*u2[i]+a23*u3[i]+
            D2dx2*(u2[i+1]-2*u2[i]+u2[i-1]);}
  }
```

Homogeneous Neumann BCs are included at i=1 and i=n as explained in Chapter 1. Only linear (Fickian) diffusion is included in eq. (3.1b).

- The coding for $u_3(x,t)$ is (with homogeneous Neumann BCs)

```
#
# u3t(x,t)
  for(i in 1:n){
    if(i==1){u3t[1]=a31*u1[1]+a32*u2[1]+
      a33*u3[1]+2*D3dx2*(u3[  2]-u3[1]);}
    if(i==n){u3t[n]=a31*u1[n]+a32*u2[n]+
      a33*u3[n]+2*D3dx2*(u3[n-1]-u3[n]);}
    if((i>1)&(i<n)){
      u3t[i]=a31*u1[i]+a32*u2[i]+a33*u3[i]+
            D3dx2*(u3[i+1]-2*u3[i]+u3[i-1]);}
  }
```

Homogeneous Neumann BCs are included at i=1 and i=n as explained in Chapter 1. Only linear (Fickian) diffusion is included in eq. (3.1c).

- The composite derivative vector ut is formed from the three derivative vectors u1t,u2t,u3t.

```
#
# Three vectors to one vector
  ut=rep(0,n3);
  for(i in 1:n){
    ut[i]    =u1t[i];
    ut[i+n]  =u2t[i];
    ut[i+n2]=u3t[i];
  }
```

- The number of calls to pde_1c is displayed as a measure of the computational effort required to compute the solution.

```
#
# Increment calls to pde_1c
  ncall <<- ncall+1;
```

- The derivative vector (LHSs of eqs. (3.1)) is returned to lsodes which requires a list. c is the R vector utility. The combination of return, list, c gives lsodes (the ODE integrator called in the main program of Listing 3.1) the required derivative vector for the next step along the solution.

```
#
# Return derivative vector
  return(list(c(ut)));
}
```

The final } concludes pde_1c.

The numerical and graphical (plotted) output is considered next.

(3.1.3) Model output

Abbreviated output for `ncase=4` follows.

[1] 6

[1] 154

t	x	u1(x,t)	u2(x,t)	u3(x,t)
0.00	0.000	0.000004	0.000000	0.000000
0.00	0.100	0.000335	0.000000	0.000000
0.00	0.200	0.011109	0.000000	0.000000
0.00	0.300	0.135335	0.000000	0.000000
0.00	0.400	0.606531	0.000000	0.000000
0.00	0.500	1.000000	0.000000	0.000000
0.00	0.600	0.606531	0.000000	0.000000
0.00	0.700	0.135335	0.000000	0.000000
0.00	0.800	0.011109	0.000000	0.000000
0.00	0.900	0.000335	0.000000	0.000000
0.00	1.000	0.000004	0.000000	0.000000

.
.
.

Output for t = 4 to 16 removed

.
.

t	x	u1(x,t)	u2(x,t)	u3(x,t)
20.00	0.000	0.000511	0.000353	-0.000590
20.00	0.100	-0.000214	-0.001228	0.000651
20.00	0.200	-0.028752	-0.023914	0.035298
20.00	0.300	-0.122689	-0.061772	0.136112

20.00	0.400	-0.175542	-0.007685	0.159998
20.00	0.500	-0.136789	0.072296	0.084966
20.00	0.600	-0.175542	-0.007685	0.159998
20.00	0.700	-0.122689	-0.061772	0.136112
20.00	0.800	-0.028752	-0.023914	0.035298
20.00	0.900	-0.000214	-0.001228	0.000651
20.00	1.000	0.000511	0.000353	-0.000590

```
ncall =  400
```

Table 3.1: Abbreviated numerical output for `ncase=4`

We can note the following details about this output.

- The values $t = 0, 4, \ldots, 20$ follow from the coding in the main program of Listing 3.1.

- The values $x = 0, 0.02, \ldots, 1$ follow from the coding in the main program of Listing 3.1, with every fifth value displayed.

- The solutions for u_1, u_2, u_3 are different since the nine elements $a_{11}, \ldots, a_{33}$ taken from [2] have nonzero, asymmetrical values (with respect to the diagonal elements a_{11}, a_{22}, a_{33}).

- Rather than the usual monotonic decay (dispersion) from just diffusion, the solutions with reaction and diffusion (linear and chemotaxis) oscillate, a manifestation of Turing oscillation, which is clear from Figs. 3.1.

- The oscillation results from complex eigenvalues (conjugate pairs) of the ODE/MOL system as discussed next.

- The solutions oscillate between positive and negative values (Figs. 3.1). Since $u_1(x, t), u_2(x, t), u_3(x, t)$ represent concentrations (of morphogens), the negative values are accommodated (explained) by considering the solutions

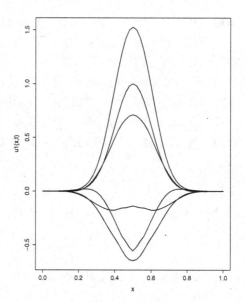

Figure 3.1a: Numerical solution of eq. (3.1a) for $u_1(x,t)$, ncase=4, from `matplot`

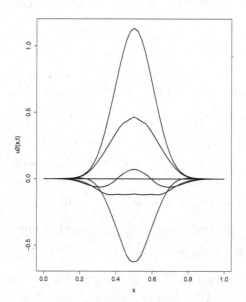

Figure 3.1b: Numerical solution of eq. (3.1b) for $u_2(x,t)$, ncase=4, from `matplot`

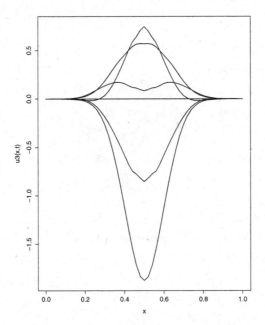

Figure 3.1c: Numerical solution of eq. (3.1c) for $u_3(x,t)$, ncase=4, from `matplot`

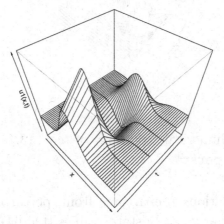

Figure 3.1d: Numerical solution of eq. (3.1a) for $u_1(x,t)$, ncase=4, from `persp`

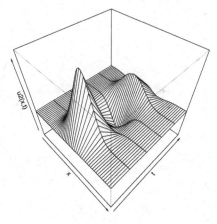

Figure 3.1e: Numerical solution of eq. (3.1b) for $u_2(x,t)$, ncase=4, from `persp`

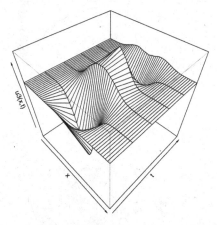

Figure 3.1f: Numerical solution of eq. (3.1c) for $u_3(x,t)$, ncase=4, from `persp`

as oscillations (departures from, perturbations around) equilibrium (steady state) values ([2], p47).

The complex (figuratively and mathematically) solutions of Figs. 3.1 give an indication of the departure from just diffusion that is possible with the RD system of PDEs (Turing models).

Figs. 3.1 indicate an oscillation in t from reaction and diffusion (linear and chemotaxis). Also, n=51 and nout=6 appear to give acceptable resolution in x and t.

Figs. 3.1 indicate the complex nature of the solutions of eqs. (3.1) for ncase=4. This can be interpreted as a form of dynamic pattern formation that would not occur with just diffusion. That is, a combination of reaction and diffusion is required, with selected values of $a_{11}, \ldots, a_{33}$ for reaction, and D_1, D_2, D_3 for diffusion.

(3.2) Eigenvalue Analysis

An eigenvalue analysis of eqs. (3.1) follows.

(3.2.1) Application to nonlinear equations

Since eqs. (3.1) are nonlinear as a consequence of the chemotaxis diffusion terms in eq. (3.1a), the following example demonstrates how an eigenvalue analysis can be applied to a nonlinear system.[3]

A main program for the eigenvalue analysis of eqs. (3.1) that is similar to the main program of Listing 2.3 follows.

(3.2.2) Main program

```
#
# 1D 3 x 3 RD
```

---

[3]Eigenvalue analysis and the associated linear algebra is defined for a linear, constant cofficient ODE system. However, an eigenvalue analysis can be applied to a nonlinear system that is linearized at a particular base point along the ODE solution. The resulting eigenvalues can be expected to change as the ODE solution evolves in t since the elements of the Jacobian matrix of the linearized ODE system change with the solution (for a linear system, the Jacobian matrix is constant).

```
#
# Delete previous workspaces
  rm(list=ls(all=TRUE))
#
# Access functions for numerical solution
  setwd("f:/turing/jacob");
  source("eigen3.R");
#
# Grid in x
  xl=0;xu=1;n=11;dx=(xu-xl)/(n-1);
  x=seq(from=xl,to=xu,by=dx);dxs=dx^2;
#
# Parameters
  ncase=4;
  if(ncase==4){
  a11=-10/3 ; a12=3      ; a13=-1  ;
  a21=-2    ; a22=7/3    ; a23=0   ;
  a31=3     ; a32=-4     ; a33=0   ;
  D1=2/3*dxs; D2=1/3*dxs; D3=0*dxs;}
  if(ncase==5){
  a11=-1   ; a12=-1  ; a13= 0  ;
  a21= 1   ; a22= 0  ; a23=-1  ;
  a31= 0   ; a32= 1  ; a33= 0  ;
  D1=1*dxs; D2=0*dxs; D3=0*dxs;}
#
# Base dependent variable vector
  n3=3*n;
  ub=rep(  0,n3);
# ub=rep(0.1,n3);
# for(i in 1:n3){
#   cat(sprintf("\n i = %2d  ub[i] = %8.4f",
#                  i,ub[i]));
# }
#
```

```
# Base dependent variable derivative vector
  utb=eigen3(t,ub,parm);
# for(i in 1:n3){
#   cat(sprintf("\n i = %2d  utb[i] = %8.4f",
# \               i,utb[i]));
# }
#
# Incremented dependent variable vector
  u=rep(0,n3);ut=rep(0,n3);
  J=matrix(0,nrow=n3,ncol=n3);
#
# Step through Jacobian matrix columns
  for(j in 1:n3){
    u[j]=ub[j]+0.01;
    ut=eigen3(t,u,parm);
#
#   Step through Jacobian matrix rows
    for(i in 1:n3){
      J[i,j]=(ut[i]-utb[i])/(u[j]-ub[j]);
#     cat(sprintf("\n i = %2d  j = %2d
#       J(i,j) = %8.4f",i,j,J[i,j]));
#
#   Next row
    }
    u[j]=ub[j];
#   cat(sprintf("\n"));
#
# Next column
  }
#
# Compute and display eigenvalues
  lam=eigen(J,only.values=TRUE);
  lamVec=lam$values;
  Re_lam=rep(0,n3);Im_lam=rep(0,n3);
```

```
  iout=0;
  for(i in 1:n3){
    Re_lam[i]=Re(lamVec[i]);Im_lam[i]=Im(lamVec[i]);
#   if(Re_lam[i]>0){
    cat(sprintf(
      "\n i = %2d  Re = %8.4f  Im = %8.4f",
      i,Re(lamVec[i]),Im(lamVec[i])));
#     iout=iout+1;
#   }
  }
  cat(sprintf("\n iout = %3d",iout));
#
# Plot eigenvalues
  Re_0=rep(0,n3);Im_0=rep(0,n3);
  Im_0[1]= -1;Im_0[n3]= 1;
# Im_0[1]=-40;Im_0[n3]=40;
  plot(Re_lam,Im_lam,lwd=2,col="black",pch="o",
       xlab="Re_lam",ylab="Im_lam",main="");
  lines(Re_0,Im_0,type="l",lwd=2);
```

Listing 3.3: Main program for eqs. (3.1)

We can note the following details about Listing 3.3.

- Previous workspaces are deleted. The **setwd** (set working directory) requires editing for the local computer to specify the directory (folder) with the R routines (note the use of / rather than the usual \). **eigen3** is a function for the $3n \times 3n$ linear, MOL system for eqs. (3.1) (discussed next).

```
  #
  # 1D 3 x 3 RD
  #
  # Delete previous workspaces
    rm(list=ls(all=TRUE))
```

```
#
# Access functions for numerical solution
  setwd("f:/turing/jacob");
  source("eigen3.R");
```

- A uniform grid in x of 11 points is defined with the `seq` utility for the interval $x_l = 0 \le x \le x_u = 1$. Therefore, the vector x has the values $x = 0, 1/10, \ldots, 1$.

```
#
# Grid in x
  xl=0;xu=1;n=11;dx=(xu-xl)/(n-1);
  x=seq(from=xl,to=xu,by=dx);dxs=dx^2;
```

The small number of grid points, $n = 11$, was selected to keep the output (if selected by deactivating comments) at a more manageable level. This main program also executes with $n = 51$ (used in Chapter 1) as discussed subsequently.

- Two cases are programmed (for the cases in [2], pp 53–54).

```
#
# Parameters
  ncase=4;
  if(ncase==4){
  a11=-10/3 ; a12=3       ; a13=-1  ;
  a21=-2    ; a22=7/3     ; a23=0   ;
  a31=3     ; a32=-4      ; a33=0   ;
  D1=2/3*dxs; D2=1/3*dxs; D3=0*dxs;}
  if(ncase==5){
  a11=-1   ; a12=-1  ; a13= 0  ;
  a21= 1   ; a22= 0  ; a23=-1  ;
  a31= 0   ; a32= 1  ; a33= 0  ;
  D1=1*dxs; D2=0*dxs; D3=0*dxs;}
```

- Base values ub for u_1, u_2, u_3 (eqs. (1.5)) are given zero values with the rep utility.

```
#
# Base dependent variable vector
  n3=3*n;
  ub=rep(  0,n3);
# ub=rep(0.1,n3);
# for(i in 1:n3){
#   cat(sprintf("\n i = %2d  ub[i] = %8.4f",
#                  i,ub[i]));
# }
```

The output is suppressed because of limited space in the discussion.

- The derivative vector utb at the base values ub is evaluated by a call to eigen3.

```
#
# Base dependent variable derivative vector
  utb=eigen3(t,ub,parm);
# for(i in 1:n3){
#   cat(sprintf("\n i = %2d  utb[i] = %8.4f",
#                  i,utb[i]));
# }
```

The output is suppressed because of limited space in the discussion.

- Vectors for f_i, u_j in eq. (2.5a) are declared (allocated).

```
#
# Incremented dependent variable vector
  u=rep(0,n3);ut=rep(0,n3);
  J=matrix(0,nrow=n3,ncol=n3);
```

A matrix for the $n \times n$ Jacobian matrix **J** in eq. (2.5b) is declared with the matrix utility.

- $u_j + \Delta u_j$ in eq. (2.5c) is defined by a change in u_j. An absolute increment 0.01 is used (u[j]=ub[j]+0.01)

```
#
# Step through Jacobian matrix columns
  for(j in 1:n3){
    u[j]=ub[j]+0.01;
    ut=eigen3(t,u,parm);
```

j is the index for columns in the Jacobian matrix and corresponds to j in eq. (2.5c). eigen3 is used to calculate the corresponding derivative vector ut.

- Within a given column (j), the rows are varied to implement eq. (2.5c). i is the index for the rows of the Jacobian matrix and corresponds to i in eq. (2.5c). The computation of the i,jth element of the Jacobian matrix is

```
#
# Step through Jacobian matrix columns
  for(j in 1:n3){
    u[j]=ub[j]+0.01;
    ut=eigen3(t,u,parm);
#
#    Step through Jacobian matrix rows
     for(i in 1:n3){
       J[i,j]=(ut[i]-utb[i])/(u[j]-ub[j]);
#      cat(sprintf("\n i = %2d  j = %2d
#        J(i,j) = %8.4f",i,j,J[i,j]));
#
#    Next row
     }
     u[j]=ub[j];
#    cat(sprintf("\n"));
#
```

```
# Next column
 }
```

The output is suppressed because of limited space in the discussion.

- At this point, all $3n \times 3n$ elements of the approximate Jacobian matrix have been computed. The $3n$ eigenvalues of **J** can now be computed (numerically, without factoring the characteristic polynomial (2.4b)).

```
#
# Compute and display eigenvalues
  lam=eigen(J,only.values=TRUE);
  lamVec=lam$values;
  Re_lam=rep(0,n3);Im_lam=rep(0,n3);
  iout=0;
  for(i in 1:n3){
    Re_lam[i]=Re(lamVec[i]);
    Im_lam[i]=Im(lamVec[i]);
    if(Re_lam[i]>0){
      cat(sprintf(
        "\n i = %2d   Re = %8.4f   Im = %8.4f",
        i,Re_lam[i],Im_lam[i]));
#     iout=iout+1;
#   }
  }
    cat(sprintf("\n iout = %3d",iout));
```

The test for unstable eigenvalues is suppressed so that all of the eigenvalues are displayed.

- The eigenvalues are plotted as points (circles, pch="o") in the complex plane. A line is added (**lines**) for a zero real part (the boundary between stable and unstable eigenvalues) using two vectors, Re_0, Im_0.

```
#
# Plot eigenvalues
  Re_0=rep(0,n3);Im_0=rep(0,n3);
  Im_0[1]= -1;Im_0[n3]= 1;
# Im_0[1]=-40;Im_0[n3]=40;
  plot(Re_lam,Im_lam,lwd=2,col="black",pch="o",
       xlab="Re_lam",ylab="Im_lam",main="");
  lines(Re_0,Im_0,type="l",lwd=2);
```

Automatic scaling of the abscissa (horizontal, x) and ordinate (vertical, y) axes accommodates variations in the plots for ncase=4,5 (Fig. 3.2 for ncase=4).

The subordinate routine eigen3 called by the main program of Listing 3.1 is considered next.

(3.2.3) Subordinate routine

The subordinate routine for the eqs. (3.1) follows.

```
  eigen3=function(t,u,parm){
#
# Function eigen3 computes the t derivative
# vector for u1(x,t), u2(x,t), u3(x,t)
#
# One vector to three vectors
  u1 =rep(0,n);u2 =rep(0,n);u3 =rep(0,n);
  u1t=rep(0,n);u2t=rep(0,n);u3t=rep(0,n);
  for(i in 1:n){
    u1[i]=u[i];
    u2[i]=u[i+n];
    u3[i]=u[i+2*n];
  }
#
# u1t(x,t)
#
```

```
# u1t(x,t)
  for(i in 1:n){
    if(i==1){u1t[1]=a11*u1[1]+a12*u2[1]+a13*u3[1];}
    if(i==n){u1t[n]=a11*u1[n]+a12*u2[n]+a13*u3[n];}
    if((i>1)&(i<n)){
    u1t[i]=a11*u1[i]+a12*u2[i]+a13*u3[i]+
      D1dx2*((u1[i]+u1[i+1])/2*(u2[i+1]-u2[i])-
            (u1[i]+u1[i-1])/2*(u2[i]-u2[i-1]))-
      D1dx2*((u1[i]+u1[i+1])/2*(u3[i+1]-u3[i])-
            (u1[i]+u1[i-1])/2*(u3[i]-u3[i-1]));}
  }
#
# u2t(x,t)
  for(i in 1:n){
    if(i==1){u2t[1]=a21*u1[1]+a22*u2[1]+a23*u3[1]+
             2*D2*(u2[  2]-u2[1])/dx^2;}
    if(i==n){u2t[n]=a21*u1[n]+a22*u2[n]+a23*u3[n]+
             2*D2*(u2[n-1]-u2[n])/dx^2;}
    if((i>1)&(i<n)){
      u2t[i]=a21*u1[i]+a22*u2[i]+a23*u3[i]+
             D2*(u2[i+1]-2*u2[i]+u2[i-1])/dx^2;}
  }
#
# u3t(x,t)
  for(i in 1:n){
    if(i==1){u3t[1]=a31*u1[1]+a32*u2[1]+a33*u3[1]+
             2*D3*(u3[  2]-u3[1])/dx^2;}
    if(i==n){u3t[n]=a31*u1[n]+a32*u2[n]+a33*u3[n]+
             2*D3*(u3[n-1]-u3[n])/dx^2;}
    if((i>1)&(i<n)){
      u3t[i]=a31*u1[i]+a32*u2[i]+a33*u3[i]+
             D3*(u3[i+1]-2*u3[i]+u3[i-1])/dx^2;}
  }
#
```

```
# Three vectors to one vector
  ut=rep(0,n3);
  for(i in 1:n){
    ut[i]    =u1t[i];
    ut[i+n]  =u2t[i];
    ut[i+2*n]=u3t[i];
  }
#
# Return derivative vector
  return(c(ut));
  }
```

Listing 3.4: Routine `eigen3` for eqs. (3.1)

`eigen3` is very similar to `pde_1c` of Listing 3.2. Only the differences are noted next.

- The function is defined. `t` is the current value of t in eqs. (3.1). `u` is the $3n = 3(11) = 33$-vector of ODE dependent variables.

  ```
  eigen3=function(t,u,parm){
  #
  # Function eigen3 computes the t derivative
  # vector for u1(x,t), u2(x,t), u3(x,t)
  ```

- The derivative vector is returned to the main program. `c` is the R utility for a numerical vector (a `list` is also used in `pde_1c` of Listing 3.2 as required by `lsodes`).

  ```
  #
  # Return derivative vector
    return(c(ut));
    }
  ```

The final `}` concludes `eigen3`.

The numerical output from the routines in Listings 3.3, 3.4 is next.

(3.2.4) Model output

For `ncase=4` in Listing 3.3, the output (with the full set of eigenvalues included) is

```
i =  1  Re =  -0.5860  Im =   1.9905
i =  2  Re =  -0.5860  Im =  -1.9905
i =  3  Re =  -0.5715  Im =   1.9769
i =  4  Re =  -0.5715  Im =  -1.9769
i =  5  Re =  -0.5297  Im =   1.9357
i =  6  Re =  -0.5297  Im =  -1.9357
i =  7  Re =  -0.4652  Im =   1.8667
i =  8  Re =  -0.4652  Im =  -1.8667
i =  9  Re =  -0.3851  Im =   1.7702
i = 10  Re =  -0.3851  Im =  -1.7702
i = 11  Re =  -0.2987  Im =   1.6482
i = 12  Re =  -0.2987  Im =  -1.6482
i = 13  Re =  -0.2156  Im =   1.5058
i = 14  Re =  -0.2156  Im =  -1.5058
i = 15  Re =  -0.1450  Im =   1.3534
i = 16  Re =  -0.1450  Im =  -1.3534
i = 17  Re =  -0.0937  Im =   1.2090
i = 18  Re =  -0.0937  Im =  -1.2090
i = 19  Re =  -1.1613  Im =   0.0000
i = 20  Re =  -1.1576  Im =   0.0000
i = 21  Re =  -1.1466  Im =   0.0000
i = 22  Re =  -1.1282  Im =   0.0000
i = 23  Re =  -1.1024  Im =   0.0000
i = 24  Re =  -0.0644  Im =   1.1002
i = 25  Re =  -0.0644  Im =  -1.1002
i = 26  Re =  -1.0693  Im =   0.0000
i = 27  Re =  -0.0552  Im =   1.0588
i = 28  Re =  -0.0552  Im =  -1.0588
i = 29  Re =  -1.0294  Im =   0.0000
i = 30  Re =  -0.9848  Im =   0.0000
```

```
i = 31  Re =  -0.9398  Im =    0.0000
i = 32  Re =  -0.9039  Im =    0.0000
i = 33  Re =  -0.8896  Im =    0.0000

iout =   0
```

Table 3.2: Numerical output for **ncase=4**, Listing 3.3

We can note the following details about this output.

- The real part of all 33 eigenvalues is negative so the PDE system of eqs. (3.1) programmed in Listings 3.1, 3.2 is stable.
- Complex eigenvalue pairs indicate the system of eqs. (3.1) is oscillatory, as demonstrated in Figs. 3.1
- **iout = 0** also indicates there are no unstable eigenvalues.

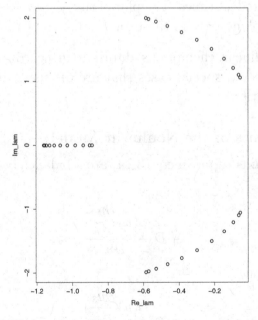

Figure 3.2: Eigenvalues for **ncase=4**

Fig. 3.2 indicates that all of the eigenvalues have negative real parts with some complex conjugate pairs as reflected in the output of Table 3.2.

In summary, the use of nonlinear chemotaxis terms in eqs. (3.1) maintained stability.

`ncase=5` in Listing 3.1 is not considered in detail to conserve space, but the reader can easily execute this case. Interesting output results.

- Eigenvalues 1 to 22 consist of the complex conjugate pair

```
Re =   -0.2151   Im =    1.3071
Re =   -0.2151   Im =   -1.3071
```

  repeated 11 times.
- Eigenvalues 23 to 33 consist of the real eigenvalue

```
Re =   -0.5698   Im =    0.0000
```

  repeated 11 times. The plot of the eigenvalues in the complex plane reflects these eigenvalues repeated 11 times.

The nonlinear chemotaxis diffusion in eq. (3.1a) produces some unexpected special cases that are discussed briefly in the following section.

(3.3) Variants of the Nonlinear Model

The chemotaxis terms in eq. (3.1a) are a balance between repulsion,

$$+D_1 \frac{\partial (u1 \frac{\partial u_2}{\partial x})}{\partial x}$$

and attraction

$$-D_1 \frac{\partial (u1 \frac{\partial u_3}{\partial x})}{\partial x}$$

The repulsion term tends to stabilize the solution (the $+$ is the usual sign of a diffusion term) while the attraction tends to destabilize the solution (from the $-$ sign). If repulsion dominates, the solution is stable while if attraction dominates, the solution can be unstable. This balance is now demonstrated for a particular case.

If the ICs for eqs. (3.1) are selected as

```
#
# ICs
  u0=rep(0,n3);
  for(i in 1:n){
    u0[i]   =c1+exp(-c2*(x[i]-0.5)^2);
    u0[i+n] =c1+exp(-c2*(x[i]-0.5)^2);
    u0[i+n2]=c1+exp(-c2*(x[i]-0.5)^2);
#   u0[i+n] =0;
#   u0[i+n2]=0;
  }
  ncall=0;
```

that is, Gaussian functions for the three ICs, the solution appears to be unstable as reflected in a convergence failure reported by lsodes. This instability apparently originates from a large gradient in u_3 in the attraction term, i.e., $\dfrac{\partial u_3}{\partial x}$.

If, however, the magnitude of the attraction term is reduced relative to the repulsion term in the ODE/MOL routine pde_1c of Listing 3.4 (note the multiplying factor 0.1),

```
u1t[i]=a11*u1[i]+a12*u2[i]+a13*u3[i]+
  D1dx2*((u1[i]+u1[i+1])/2*(u2[i+1]-u2[i])-
         (u1[i]+u1[i-1])/2*(u2[i]-u2[i-1]))-
  0.1*D1dx2*((u1[i]+u1[i+1])/2*(u3[i+1]-u3[i])-
             (u1[i]+u1[i-1])/2*(u3[i]-u3[i-1]));}
```

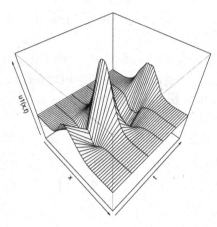

Figure 3.3a: Numerical solution of eq. (3.1a) for $u_1(x,t)$, `ncase=4`, reduced attraction

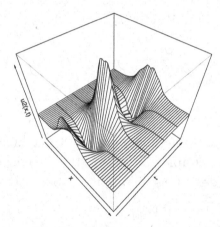

Figure 3.3b: Numerical solution of eq. (3.1a) for $u_2(x,t)$, `ncase=4`, reduced attraction

the solution is stable and oscillatory. This solution is displayed in 3D in Figs. 3.3.

This example demonstrates the complexities of a nonlinear system based on chemotaxis in eq. (3.1a). Experience has indicated if repulsion dominates, the solution is stable, while if attraction dominates, the solution is unstable. The stability

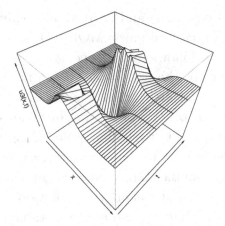

Figure 3.3c: Numerical solution of eq. (3.1a) for $u_3(x,t)$, ncase=4, reduced attraction

of attraction-repulsion systems has been studied analytically [1, 3], with emphasis on the relative magnitudes of attraction and repulsion.

(3.4) Summary and Conclusions

Computer-based experimentation to investigate the features of a nonlinear system is straightforward within the MOL framework. This type of investigation can be enhanced by computing and displaying the PDE RHS terms (typically spatial derivatives and source terms) and the LHS terms (usually derivatives in time). In this way, the contributions of the individual PDE terms can be observed and evaluated. This insight may in turn suggest modifications of the model, e.g., addition or removal of terms, additional PDEs, with, ideally, convergence to a useful mathematical model that might, for example, achieve agreement with reported experimental data or observed features/phenomena.

Also, we should realize that eigenvalue analysis as described in the preceding examples has limited utility for an understanding of the stability of nonlinear systems. In other words, the linear algebra of eqs. (2.4) and (2.5) applies only to the case of linear PDEs and therefore linear MOL ODEs.

This conclusion follows from the constant Jacobian matrix of linear ODEs, but not the variable Jacobian matrix of nonlinear ODEs, i.e., for the latter, the Jacobian matrix will vary as the solution evolves. In other words, the Jacobian matrix can be considered to apply at a particular point along the solution of nonlinear ODEs, and as it changes, the associated eigenvalues will change. For example, the eigenvalues may all have nonpositive real parts at a point along the solution (Table 3.2, Fig. 3.2) and the MOL ODE system appears stable (Figs. 3.1, 3.3). Then some of the eigenvalues may assume positive real parts at another point along the solution.

This can be demonstrated with the preceding chemotaxis model by changing the base values of the dependent variables (which could represent a change in the Jacobian matrix as the solution proceeds). For example, if the base values of the dependent variables are changed (in Listing 3.3) from

```
ub=rep(  0,n3);
```

for which all of the eigenvalues have nonpositive real parts (Table 3.2, Fig. 3.2) to

```
ub=rep(0.1,n3);
```

some of the eigenvalues have positive real parts, which infers that the solution is unstable. But this is not observed in Figs. 3.1, 3.3, where the solution remains bounded for $0 \leq t \leq 20$.

Therefore, the eigenvalue analysis as a means of studying the stability of the nonlinear PDE system of eqs. (3.1) should be used with reservation. Ultimately, what matters is whether the solution is stable or unstable, as observed when it is computed. The preceding eigenvalue analysis was included to reflect the discussion of eigenvalues in [2] that is primarily for linear reaction-diffusion PDEs systems. For the application to nonlinear PDEs, it is essential to understand the limitations of the eigenvalue approach to the analysis of nonlinear stability.

References

[1] Lin, K., C. Mu, and Y. Gao (2016), Boundedness and blow up in the higher-dimensional attraction−repulsion chemotaxis system with nonlinear diffusion, *Journal of Differential Equations*, **261**, no. 8, pp 4524-4572

[2] Turing, A.M. (1952), The chemical basis of morphogenesis, *Philosophical Transactions of the Royal Society of London, Series B, Biological Sciences*, **237**, no. 641

[3] Yu, H., Q. Guo and S. Zheng (2017), Finite time blow-up of nonradial solutions in an attraction−repulsion chemotaxis system, *Nonlinear Analysis: Real World Applications*, **34**, pp 335-342S

Chapter 4

Alternate Coordinate Systems

The method of lines (MOL) solution discussed in Chapter 1, and the eigenvalue analysis discussed in Chapter 2 pertain to a linear reaction-diffusion (RD) partial differential equation (PDE) model. A central result is the oscillation that is possible with a linear RD system, and which raises the question whether oscillation also occurs in other coordinate systems. The following discussion confirms that oscillation occurs in 1D cylindrical and spherical coordinates.

(4.1) 1D 3 × 3 Linear Model in Cylindrical Coordinates

The 3 × 3 linear model of eqs. (1.5) is first written in cylindrical coordinates (r, θ, z) with r as the spatial variable (for a 1D model).

$$\frac{\partial u_1}{\partial t} = a_{11}u_1 + a_{12}u_2 + a_{13}u_3 + D_1 \left(\frac{\partial^2 u_1}{\partial r^2} + \frac{1}{r}\frac{\partial u_1}{\partial r} \right) \qquad (4.1a)$$

$$\frac{\partial u_2}{\partial t} = a_{21}u_1 + a_{22}u_2 + a_{23}u_3 + D_2 \left(\frac{\partial^2 u_2}{\partial r^2} + \frac{1}{r}\frac{\partial u_2}{\partial r} \right) \qquad (4.1b)$$

$$\frac{\partial u_3}{\partial t} = a_{31}u_1 + a_{32}u_2 + a_{43}u_3 + D_3 \left(\frac{\partial^2 u_3}{\partial r^2} + \frac{1}{r}\frac{\partial u_3}{\partial r} \right) \qquad (4.1c)$$

Eqs. (4.1) are first order in t and second order in r. They therefore each require one initial condition (IC) and two boundary conditions (BCs).

$$u_1(r, t = 0) = f_1(r); \ u_2(r, t = 0) = f_2(r); \ u_3(r, t = 0) = f_3(r)$$
$$(4.2\text{a,b,c})$$

where $f_1(r), f_2(r), f_3(r)$ are functions to be specified.

Homogeneous Neumann BCs are specified.

$$\frac{\partial u_1(r = 0, t)}{\partial r} = \frac{\partial u_2(r = 0, t)}{\partial r} = \frac{\partial u_3(r = 0, t)}{\partial r} = 0 \quad (4.3\text{a,b,c})$$

$$\frac{\partial u_1(r = r_0, t)}{\partial r} = \frac{\partial u_2(r = r_0, t)}{\partial r} = \frac{\partial u_3(r = r_0, t)}{\partial r} = 0$$
$$(4.3\text{d,e,f})$$

BCs (4.3a,b,c) express symmetry at $r = 0$. Eqs. (4.3d,e,f) are *zero flux* or *impermeable* BCs.

The MOL solution of eqs. (4.1), (4.2) and (4.3) is considered next.

(4.1.1) Main program

The main program for eqs. (4.1), (4.2) and (4.3) is similar to the main program of Listing 1.1.

```
#
# 1D 3 x 3 Turing, cylindrical
#
# Delete previous workspaces
  rm(list=ls(all=TRUE))
#
# Access ODE integrator
  library("deSolve");
#
# Access functions for numerical solution
  setwd("f:/turing/1D");
```

```
  source("pde_1d.R");
#
# Grid in r
  rl=0;ru=1;n=51;dr=(ru-rl)/(n-1);
  r=seq(from=rl,to=ru,by=dr);
  n2=2*n;n3=3*n;
#
# Parameters
  ncase=4;
  if(ncase==1){
    a11  =0; a12  =0; a13  =0;
    a21  =0; a22  =0; a23  =0;
    a31  =0; a32  =0; a33  =0;
    D1=dr^2; D2=dr^2; D3=dr^2;
    t0   =0; tf  =20; nout =6;
    c1   =0; c2  =50;
    zl   =0; zu   =1;}
  if(ncase==2){
    a11= -1; a12  =0; a13  =0;
    a21 = 0; a22 =-1; a23 = 0;
    a31 = 0; a32  =0; a33 =-1;
    D1=dr^2; D2=dr^2; D3=dr^2;
    t0   =0; tf  =20; nout =6;
    c1   =0; c2  =50;
    zl   =0; zu   =1;}
  if(ncase==3){
    a11  =1; a12  =0; a13  =0;
    a21  =0; a22  =1; a23  =0;
    a31  =0; a32  =0; a33  =1;
    D1=dr^2; D2=dr^2; D3=dr^2;
    t0   =0; tf   =2; nout =6;
    c1   =0; c2  =50;
    zl   =0; zu   =1;}
  if(ncase==4){
```

```
    a11   =-10/3; a12      =3; a13   =-1;
    a21     =-2; a22    =7/3; a23    =0;
    a31      =3; a32     =-4; a33    =0;
    D1=2/3*dr^2; D2=1/3*dr^2; D3=0*dr^2;
    t0      =0; tf     =20; nout   =6;
    c1      =0; c2     =50;
    zl     =-2; zu      =3;}
    if(ncase==5){
    a11    =-1; a12    =-1; a13   = 0;
    a21     =1; a22     =0; a23   =-1;
    a31     =0; a32     =1; a33    =0;
    D1=1*dr^2; D2=0*dr^2; D3=0*dr^2;
    t0      =0; tf     =20; nout   =6;
    c1      =0; c2     =50;
    zl     =-1; zu      =2;}
#
# Factors used in pde_1h.R
    D1dr2=D1/dr^2  ;D2dr2=D2/dr^2  ;
    D3dr2=D3/dr^2  ;
    D12dr=D1/(2*dr);D22dr=D2/(2*dr);
    D32dr=D3/(2*dr);
#
# Independent variable for ODE integration
    tout=seq(from=t0,to=tf,by=(tf-t0)/(nout-1));
#
# ICs
    u0=rep(0,n3);
    for(i in 1:n){
      u0[i]   =c1+exp(-c2*(r[i]-0.5)^2);
      u0[i+n] =c1+exp(-c2*(r[i]-0.5)^2);
      u0[i+n2]=c1+exp(-c2*(r[i]-0.5)^2);
#     u0[i+n] =0;
#     u0[i+n2]=0;
    }
```

```
  ncall=0;
#
# ODE integration
  out=lsodes(y=u0,times=tout,func=pde_1d,
      sparsetype="sparseint",rtol=1e-6,
      atol=1e-6,maxord=5);
  nrow(out)
  ncol(out)
#
# Arrays for numerical solution
  u1=matrix(0,nrow=n,ncol=nout);
  u2=matrix(0,nrow=n,ncol=nout);
  u3=matrix(0,nrow=n,ncol=nout);
  t=rep(0,nout);
  for(it in 1:nout){
  for(i  in 1:n){
    u1[i,it]=out[it,i+1];
    u2[i,it]=out[it,i+1+n];
    u3[i,it]=out[it,i+1+n2];
      t[it]=out[it,1];
  }
  }
#
# Display selected output
  for(it in 1:nout){
    cat(sprintf("\n     t         r    u1(r,t)
            u2(r,t)    u3(r,t)\n"));
    iv=seq(from=1,to=n,by=5);
    for(i in iv){
      cat(sprintf(
        "%6.2f%9.3f%10.6f%10.6f%10.6f\n",
        t[it],r[i],u1[i,it],u2[i,it],u3[i,it]));
      }
      cat(sprintf("\n"));
```

```
    }
    cat(sprintf(" ncall = %4d\n",ncall));
#
# Plot 2D numerical solution
    matplot(r,u1,type="1",lwd=2,col="black",
        lty=1,xlab="r",ylab="u1(r,t)",main="");
    matplot(r,u2,type="1",lwd=2,col="black",
        lty=1,xlab="r",ylab="u2(r,t)",main="");
    matplot(r,u3,type="1",lwd=2,col="black",
        lty=1,xlab="r",ylab="u3(r,t)",main="");
#
# Plot 3D numerical solution
    persp(r,t,u1,theta=45,phi=45,xlim=c(rl,ru),
            ylim=c(t0,tf),xlab="r",ylab="t",
            zlab="u1(r,t)");
    persp(r,t,u2,theta=45,phi=45,xlim=c(rl,ru),
            ylim=c(t0,tf),xlab="r",ylab="t",
            zlab="u2(r,t)");
    persp(r,t,u3,theta=45,phi=45,xlim=c(rl,ru),
            ylim=c(t0,tf),xlab="r",ylab="t",
            zlab="u3(r,t)");
```

Listing 4.1: Main program for eqs. (4.1),(4.2),(4.3)

We can note the following details about Listing 4.1.

- Previous workspaces are removed. Then the ODE integrator library **deSolve** is accessed. The **setwd** (set working directory) uses / rather than the usual \, and requires editing on the local computer (to specify the directory with the R files).

```
#
# 1D 3 x 3 Turing, cylindrical
#
```

```
# Delete previous workspaces
  rm(list=ls(all=TRUE))
#
# Access ODE integrator
  library("deSolve");
#
# Access functions for numerical solution
  setwd("f:/turing/1D");
  source("pde_1d.R");
```

pde_1d is the routine for the MOL ODEs (discussed subsequently).

- A uniform grid in r of 51 points is defined with the seq utility for the interval $r_l = 0 \le r \le r_u = 1$. Therefore, the vector r has the values $r = 0, 1/50, \ldots, 1$.

```
#
# Grid in r
  rl=0;ru=1;n=51;dr=(ru-rl)/(n-1);
  r=seq(from=rl,to=ru,by=dr);
  n2=2*n;n3=3*n;
```

- Five cases are programmed with variations in the model parameters, i.e., ncase=1,2,3,4,5.

```
#
# Parameters
  ncase=4;
  if(ncase==1){
    a11  =0; a12  =0; a13  =0;
    a21  =0; a22  =0; a23  =0;
    a31  =0; a32  =0; a33  =0;
    D1=dr^2; D2=dr^2; D3=dr^2;
    t0   =0; tf   =20; nout =6;
    c1   =0; c2   =50;
```

```
    zl   =0; zu   =1;}
if(ncase==2){
    a11= -1; a12  =0; a13  =0;
    a21 = 0; a22 =-1; a23 = 0;
    a31 = 0; a32  =0; a33 =-1;
    D1=dr^2; D2=dr^2; D3=dr^2;
    t0   =0; tf   =20; nout =6;
    c1   =0; c2  =50;
    zl   =0; zu   =1;}
if(ncase==3){
    a11  =1; a12  =0; a13  =0;
    a21  =0; a22  =1; a23  =0;
    a31  =0; a32  =0; a33  =1;
    D1=dr^2; D2=dr^2; D3=dr^2;
    t0   =0; tf   =2; nout =6;
    c1   =0; c2  =50;
    zl   =0; zu   =1;}
if(ncase==4){
a11  =-10/3; a12      =3; a13      =-1;
a21       =-2; a22     =7/3; a23      =0;
a31       =3; a32       =-4; a33      =0;
D1=2/3*dr^2; D2=1/3*dr^2; D3=0*dr^2;
t0        =0; tf        =20; nout    =6;
c1        =0; c2        =50;
zl        =-2; zu        =3;}
if(ncase==5){
a11   =-1; a12   =-1; a13   = 0;
a21   =1; a22   =0; a23   =-1;
a31   =0; a32   =1; a33   =0;
D1=1*dr^2; D2=0*dr^2; D3=0*dr^2;
t0   =0; tf   =20; nout   =6;
c1   =0; c2  =50;
zl   =-1; zu   =2;}
```

As in the case of 1D Cartesian coordinates in Chapter 1, we look to determine if the solutions oscillate for ncase=4,5 (that is, if some of the eigenvalues of the MOL/ODEs are complex conjugate pairs).

- ncase==4: The reaction constants $a_{11}, \ldots, a_{33}$ and diffusivities are taken from Turing ([2], p53, eqs. (8.5)).
- ncase==5: The reaction constants $a_{11}, \ldots, a_{33}$ and diffusivities are taken from Turing ([2], p54, eqs. (8.7)).

This is a marked distinction from Fickian diffusion which disperses the solution monotonically in t.

- Factors used in the ODE/MOL routine pde_1d are defined.

```
#
# Factors used in pde_1d.R
  D1dr2=D1/dr^2  ;D2dr2=D2/dr^2  ;
  D3dr2=D3/dr^2  ;
  D12dr=D1/(2*dr);D22dr=D2/(2*dr);
  D32dr=D3/(2*dr);
```

Each term includes a division by $\Delta r^2 = dr^2$ (for the second derivative in r in eqs. (4.1)) or $2\Delta r = 2dr$ (for the first derivative in r in eqs. (4.1)).

- A uniform grid in t of 6 output points is defined with the seq utility for the interval $t_0 = 0 \le t \le t_f = 20$. Therefore, the vector tout has the values $t = 0, 4, \ldots, 20$.

```
#
# Independent variable for ODE integration
  tout=seq(from=t0,to=tf,by=(tf-t0)/(nout-1));
```

(t0,tf defined previously as parameters).

At this point, the intervals of r and t in eqs. (4.1) are defined.

- IC functions in eqs. (4.2) are defined. n2,n3 are constants defined previously.

```
#
# ICs
  u0=rep(0,n3);
  for(i in 1:n){
    u0[i]    =c1+exp(-c2*(r[i]-0.5)^2);
    u0[i+n]  =c1+exp(-c2*(r[i]-0.5)^2);
    u0[i+n2]=c1+exp(-c2*(r[i]-0.5)^2);
#   u0[i+n]  =0;
#   u0[i+n2]=0;
  }
  ncall=0;
```

The functions for u_2, u_3 can also be activated (uncommented) to implement zero functions.

Another possibility would be to use randomly distributed IC values to investigate patterning in the solutions, but this might require an increase in the number of grid points in r (above n=51) for improved spatial resolution. The counter for the calls to pde_1d is initialized (and passed to pde_1d without a special designation).

- The system of 3(51) = 153 MOL/ODEs is integrated by the library integrator lsodes (available in deSolve) with the sparse matrix option specified. As expected, the inputs to lsodes are the ODE function, pde_1d, the IC vector u0, and the vector of output values of t, tout. The length of u0 (e.g., 153) informs lsodes how many ODEs are to be integrated. func,y,times are reserved names.

```
#
# ODE integration
```

```
out=lsodes(y=u0,times=tout,func=pde_1d,
    sparsetype="sparseint",rtol=1e-6,
    atol=1e-6,maxord=5);
nrow(out)
ncol(out)
```

The numerical solution to the ODEs is returned in matrix out. In this case, out has the dimensions $nout \times (n+1) = 6 \times 154$. The offset $n+1$ is required since the first element of each column has the output t (also in tout), and the $2, \ldots, n+1 = 2, \ldots, 154$ column elements have the 153 ODE solutions. This indexing of out in used next.

- The ODE solution is placed in 3 51×6 matrices, u1,u2,u3, for subsequent plotting (by stepping through the solution with respect to r and t within a pair of fors).

```
#
# Arrays for numerical solution
  u1=matrix(0,nrow=n,ncol=nout);
  u2=matrix(0,nrow=n,ncol=nout);
  u3=matrix(0,nrow=n,ncol=nout);
  t=rep(0,nout);
  for(it in 1:nout){
  for(i  in 1:n){
   u1[i,it]=out[it,i+1];
   u2[i,it]=out[it,i+1+n];
   u3[i,it]=out[it,i+1+n2];
     t[it]=out[it,1];
  }
  }
```

- The numerical solutions are displayed.

```
#
# Display selected output
  for(it in 1:nout){
```

```
      cat(sprintf("\n      t          r    u1(r,t)
          u2(r,t)    u3(r,t)\n"));
      iv=seq(from=1,to=n,by=5);
      for(i in iv){
          cat(sprintf(
        "%6.2f%9.3f%10.6f%10.6f%10.6f\n",
        t[it],r[i],u1[i,it],u2[i,it],u3[i,it]));
        }
        cat(sprintf("\n"));
      }
    cat(sprintf(" ncall = %4d\n",ncall));
```

To conserve space, only every fifth value in r of the solutions is displayed numerically (using the subscript iv).

- The solutions to eqs. (4.1), $u_1(r,t), u_2(r,t), u_3(r,t)$, are plotted vs r with t as a parameter in 2D with matplot.

```
#
# Plot 2D numerical solution
  matplot(r,u1,type="l",lwd=2,col="black",
    lty=1,xlab="r",ylab="u1(r,t)",main="");
  matplot(r,u2,type="l",lwd=2,col="black",
    lty=1,xlab="r",ylab="u2(r,t)",main="");
  matplot(r,u3,type="l",lwd=2,col="black",
    lty=1,xlab="r",ylab="u3(r,t)",main="");
```

Note that rows of r (nrows=n=51) equals the rows of u1,u2,u3.

- The solutions to eqs. (4.1), $u_1(r,t), u_2(r,t), u_3(r,t)$, are plotted vs r and t in 3D perspective with persp.

```
#
# Plot 3D numerical solution
  persp(r,t,u1,theta=45,phi=45,xlim=c(rl,ru),
      ylim=c(t0,tf),xlab="r",ylab="t",
      zlab="u1(r,t)");
```

```
persp(r,t,u2,theta=45,phi=45,xlim=c(rl,ru),
    ylim=c(t0,tf),xlab="r",ylab="t",
    zlab="u2(r,t)");
persp(r,t,u3,theta=45,phi=45,xlim=c(rl,ru),
    ylim=c(t0,tf),xlab="r",ylab="t",
    zlab="u3(r,t)");
```

Automatic scaling in z is used. If the three plots are to have a common vertical scale in z, zlim=c(zl,zu) defined previously could be used. r,t have dimensions in agreement with u1,u2,u3 (nrows=n=51, ncols=nout=6).

The ODE/MOL routine called in the main program of Listing 4.1, pde_1d, follows.

(4.1.2) Subordinate routine

The ODE/MOL routine for eqs. (4.1) is listed next.

```
  pde_1d=function(t,u,parm){
#
# Function pde_1d computes the t derivative
# vector for u1(r,t), u2(r,t), u3(r,t)
#
# One vector to three vectors
  u1 =rep(0,n);u2 =rep(0,n);u3 =rep(0,n);
  u1t=rep(0,n);u2t=rep(0,n);u3t=rep(0,n);
  for(i in 1:n){
    u1[i]=u[i];
    u2[i]=u[i+n];
    u3[i]=u[i+n2];
  }
#
# u1t(r,t)
  for(i in 1:n){
```

```
    if(i==1){u1t[1]=a11*u1[1]+a12*u2[1]+
      a13*u3[1]+4*D1dr2*(u1[  2]-u1[1]);}
    if(i==n){u1t[n]=a11*u1[n]+a12*u2[n]+
      a13*u3[n]+2*D1dr2*(u1[n-1]-u1[n]);}
    if((i>1)&(i<n)){
      u1t[i]=a11*u1[i]+a12*u2[i]+a13*u3[i]+
            D1dr2*(u1[i+1]-2*u1[i]+u1[i-1])+
            D12dr*(1/r[i])*(u1[i+1]-u1[i-1]);}
    }
#
# u2t(r,t)
  for(i in 1:n){
    if(i==1){u2t[1]=a21*u1[1]+a22*u2[1]+
      a23*u3[1]+4*D2dr2*(u2[  2]-u2[1]);}
    if(i==n){u2t[n]=a21*u1[n]+a22*u2[n]+
      a23*u3[n]+2*D2dr2*(u2[n-1]-u2[n]);}
    if((i>1)&(i<n)){
      u2t[i]=a21*u1[i]+a22*u2[i]+a23*u3[i]+
            D2dr2*(u2[i+1]-2*u2[i]+u2[i-1])+
            D22dr*(1/r[i])*(u2[i+1]-u2[i-1]);}
    }
#
# u3t(r,t)
  for(i in 1:n){
    if(i==1){u3t[1]=a31*u1[1]+a32*u2[1]+
      a33*u3[1]+4*D3dr2*(u3[  2]-u3[1]);}
    if(i==n){u3t[n]=a31*u1[n]+a32*u2[n]+
      a33*u3[n]+2*D3dr2*(u3[n-1]-u3[n]);}
    if((i>1)&(i<n)){
      u3t[i]=a31*u1[i]+a32*u2[i]+a33*u3[i]+
            D3dr2*(u3[i+1]-2*u3[i]+u3[i-1])+
            D32dr*(1/r[i])*(u3[i+1]-u3[i-1]);}
    }
#
```

```
# Three vectors to one vector
  ut=rep(0,n3);
  for(i in 1:n){
    ut[i]    =u1t[i];
    ut[i+n]  =u2t[i];
    ut[i+n2]=u3t[i];
  }
#
# Increment calls to pde_1d
  ncall <<- ncall+1;
#
# Return derivative vector
  return(list(c(ut)));
  }
```

Listing 4.2: ODE/MOL routine pde_1d for eqs. (4.1)

We can note the following details about pde_1d.

- The function is defined.

  ```
  pde_1d=function(t,u,parm){
  #
  # Function pde_1d computes the t derivative
  # vector for u1(r,t), u2(r,t), u3(r,t)
  ```

  t is the current value of t in eqs. (4.1). u is the 153-vector of ODE/MOL dependent variables. parm is an argument to pass parameters to pde_1d (unused, but required in the argument list). The arguments must be listed in the order stated to properly interface with lsodes called in the main program of Listing 4.1. The composite derivative vector of the LHSs of eqs. (4.1) is calculated next and returned to lsodes.

- The dependent variable vectors are placed in three vectors to facilitate the programming of eqs. (4.1).

```
#
# One vector to three vectors
  u1 =rep(0,n);u2 =rep(0,n);u3 =rep(0,n);
  u1t=rep(0,n);u2t=rep(0,n);u3t=rep(0,n);
  for(i in 1:n){
    u1[i]=u[i];
    u2[i]=u[i+n];
    u3[i]=u[i+n2];
  }
```

Vectors are also defined for the LHS derivatives in t of eqs. (4.1).

- $\dfrac{\partial u_1}{\partial t}$ of eq. (4.1a) is programmed in a for that steps through x, for(i in 1:n).

```
#
# u1t(r,t)
  for(i in 1:n){
    if(i==1){u1t[1]=a11*u1[1]+a12*u2[1]+
      a13*u3[1]+4*D1dr2*(u1[  2]-u1[1]));}
    if(i==n){u1t[n]=a11*u1[n]+a12*u2[n]+
      a13*u3[n]+2*D1dr2*(u1[n-1]-u1[n]));}
    if((i>1)&(i<n)){
      u1t[i]=a11*u1[i]+a12*u2[i]+a13*u3[i]+
             D1dr2*(u1[i+1]-2*u1[i]+u1[i-1])+
             D12dr*(1/r[i])*(u1[i+1]-u1[i-1]));}
  }
```

This coding requires some additional explanation.

- At $r = r_l = 0$ ($i = 1$), the homogeneous Neumann BCs (4.3a,b,c) are used. For the diffusion term in eq. (4.1a),

$$D_1 \left(\frac{\partial^2 u_1(r=0,t)}{\partial r^2} + \frac{1}{r} \frac{\partial u_1(r=0,t)}{\partial r} \partial r \right)$$

$$= D_1 2 \frac{\partial^2 u_1(r=0,t)}{\partial r^2} \approx$$

$$D_1 2 \frac{u_1(r=\Delta r,t)}{\Delta r^2}$$

$$-2 D_1 2 \frac{u_1(r=0,t)}{\Delta r^2}$$

$$+ D_1 2 \frac{u_1(r=-\Delta r,t)}{\Delta r^2}$$

$$= D_1 4 \left(\frac{u_1(r=\Delta r,t) - u_1(r=0,t)}{\Delta r^2} \right)$$

with the coding

`4*D1dr2*(u1[  2]-u1[1])`

The final form for the radial group in eq. (4.1a) at $r = 0$ follows from the indeterminate form $\dfrac{1}{r} \dfrac{\partial u_1(r,t)}{\partial r}$ at $r = 0$ as regularized by *l'Hospital's* rule [1], p331,

$$\frac{\partial^2 u_1(r=0,t)}{\partial r^2} + \frac{1}{r} \frac{\partial u_1(r=0,t)}{\partial r} \partial r = 2 \frac{\partial^2 u_1(r=0,t)}{\partial r^2}$$

and the fictitious value $u_1(r = -\Delta r, t)$ is replaced through BC (4.3a)

$$\frac{\partial u_1(r=0,t)}{\partial r} \approx \frac{u_1(r=\Delta r,t) - u_1(r=-\Delta r,t)}{2\Delta r} = 0$$

or

$$u_1(r = -\Delta r, t) = u_1(r = \Delta r, t)$$

– The same reasoning applies to the application of homogeneous Neumann BCs (4.3,d,e,f) at the right boundary $r = r_0 = 1$ $(i = n)$. The indeterminate form does not occur since $r_0 = 1$ (not zero) and therefore the radial group in eq. (4.1a) is

$$D_1 \left(\frac{\partial^2 u_1(r = r_0, t)}{\partial r^2} + \frac{1}{r_0} \frac{\partial u_1(r = r_0, t)}{\partial r} \partial r \right)$$

$$= D_1 \frac{\partial^2 u_1(r = r_0, t)}{\partial r^2}$$

$$\approx D_1 \frac{u_1(r = r_0 + \Delta r, t)}{\Delta r^2}$$

$$-2D_1 \frac{u_1(r = r_0, t)}{\Delta r^2}$$

$$+D_1 \frac{u_1(r = r_0 - \Delta r, t)}{\Delta r^2}$$

$$= 2D_1 \left(\frac{u_1(r = r_0 - \Delta r, t) - u_1(r = r_0, t)}{\Delta r^2} \right)$$

with the coding

`2*D1dr2*(u1[n-1]-u1[n])`

A FD approximation of BC (4.3d) was used in the radial group in eq. (4.1a) at $r = r_0$ to eliminate the fictitious value $u_1(r = r_0 + \Delta r, t)$

$$u_1(r = r_0 + \Delta r, t) \approx u_1(r = r_0 - \Delta r, t)$$

- For the interior points $r_l + \Delta r \leq r \leq r_u - \Delta r$, the radial group in eq. (4.1a) is approximated with FDs.

$$D_1 \left(\frac{\partial^2 u_1}{\partial r^2} + \frac{1}{r} \frac{\partial u_1}{\partial r} \right)$$

$$\approx D_1 \left(\frac{u_1(r + \Delta r, t) - 2u_1(r, t) + u_1(r - \Delta r, t)}{\Delta r^2} \right.$$

$$\left. + \frac{1}{r} \frac{u_1(r + \Delta r, t) - u_1(r - \Delta r, t)}{2\Delta r} \right)$$

The corresponding coding is

```
D1dr2*(u1[i+1]-2*u1[i]+u1[i-1])+
D12dr*(1/r[i])*(u1[i+1]-u1[i-1])
```

- The coding for $u_2(r,t)$ (with homogeneous Neumann BCs) is

```
#
# u2t(r,t)
  for(i in 1:n){
    if(i==1){u2t[1]=a21*u1[1]+a22*u2[1]+
    a23*u3[1]+4*D2dr2*(u2[  2]-u2[1]);}
    if(i==n){u2t[n]=a21*u1[n]+a22*u2[n]+
    a23*u3[n]+2*D2dr2*(u2[n-1]-u2[n]);}
    if((i>1)&(i<n)){
      u2t[i]=a21*u1[i]+a22*u2[i]+a23*u3[i]+
      D2dr2*(u2[i+1]-2*u2[i]+u2[i-1])+
      D22dr*(1/r[i])*(u2[i+1]-u2[i-1]);}
  }
```

Homogeneous Neumann BCs are included at i=1 and i=n as explained for $u_1(r,t)$.

- The coding for $u_3(r,t)$ is (with homogeneous Neumann BCs)

```
#
# u3t(r,t)
  for(i in 1:n){
    if(i==1){u3t[1]=a31*u1[1]+a32*u2[1]+
    a33*u3[1]+4*D3dr2*(u3[  2]-u3[1]);}
    if(i==n){u3t[n]=a31*u1[n]+a32*u2[n]+
    a33*u3[n]+2*D3dr2*(u3[n-1]-u3[n]);}
    if((i>1)&(i<n)){
      u3t[i]=a31*u1[i]+a32*u2[i]+a33*u3[i]+
      D3dr2*(u3[i+1]-2*u3[i]+u3[i-1])+
```

```
                       D32dr*(1/r[i])*(u3[i+1]-u3[i-1]);}
  }
```

Homogeneous Neumann BCs are included at i=1 and i=n as explained for $u_1(r,t)$.

- The composite derivative vector ut is formed from the three derivative vectors u1t,u2t,u3t.

```
#
# Three vectors to one vector
  ut=rep(0,n3);
  for(i in 1:n){
    ut[i]   =u1t[i];
    ut[i+n] =u2t[i];
    ut[i+n2]=u3t[i];
  }
```

- The number of calls to pde_1d is displayed as a measure of the computational effort required to compute the solution.

```
#
# Increment calls to pde_1d
  ncall <<- ncall+1;
```

- The derivative vector (LHSs of eqs. (4.1)) is returned to lsodes which requires a list. c is the R vector utility. The combination of return, list, c gives lsodes (the ODE integrator called in the main program of Listing 4.1) the required derivative vector for the next step along the solution.

```
#
# Return derivative vector
  return(list(c(ut)));
  }
```

The final } concludes pde_1d.

The numerical and graphical (plotted) output is considered next.

(4.1.3) Model output

Abbreviated output for ncase=4 follows.

[1] 6

[1] 154

t	r	u1(r,t)	u2(r,t)	u3(r,t)
0.00	0.000	0.000004	0.000004	0.000004
0.00	0.100	0.000335	0.000335	0.000335
0.00	0.200	0.011109	0.011109	0.011109
0.00	0.300	0.135335	0.135335	0.135335
0.00	0.400	0.606531	0.606531	0.606531
0.00	0.500	1.000000	1.000000	1.000000
0.00	0.600	0.606531	0.606531	0.606531
0.00	0.700	0.135335	0.135335	0.135335
0.00	0.800	0.011109	0.011109	0.011109
0.00	0.900	0.000335	0.000335	0.000335
0.00	1.000	0.000004	0.000004	0.000004

.
.
.   .
.
.

Output for t = 4 to 16 removed

.
.
.   .
.
.

t	r	u1(r,t)	u2(r,t)	u3(r,t)
20.00	0.000	-0.000105	-0.000058	0.000107
20.00	0.100	-0.000543	-0.000559	0.000878
20.00	0.200	0.005538	-0.000498	-0.002619
20.00	0.300	0.139275	0.083179	-0.153238

```
20.00    0.400   0.718562   0.549302  -0.867038
20.00    0.500   1.224187   0.998045  -1.508549
20.00    0.600   0.727110   0.568487  -0.884024
20.00    0.700   0.146151   0.094883  -0.166076
20.00    0.800   0.007814   0.002060  -0.006327
20.00    0.900  -0.000176  -0.000293   0.000401
20.00    1.000  -0.000034  -0.000024   0.000040
```

```
ncall =   363
```

Table 4.1: Abbreviated numerical output for `ncase=4`

We can note the following details about this output.

- The values $t = 0, 4, \ldots, 20$ follow from the coding in the main program of Listing 4.1.
- The values $r = 0, 0.02, \ldots, 1$ follow from the coding in the main program of Listing 4.1, with every fifth value displayed.
- The solutions for u_1, u_2, u_3 are different since the nine elements $a_{11}, \ldots, a_{33}$ taken from [2] have nonzero, asymmetrical values (with respect to the diagonal elements a_{11}, a_{22}, a_{33}).
- Rather than usual monotonic decay (dispersion) from just diffusion, the solutions with reaction and diffusion (linear and chemotaxis) oscillate, a manifestation of Turing oscillation, which is clear from Figs. 4.1.
- The oscillation results from complex eigenvalues (conjugate pairs) of the ODE/MOL system as discussed next.
- The solutions oscillate between positive and negative values (Figs. 4.1). Since $u_1(r, t), u_2(r, t), u_3(r, t)$ represent concentrations (of morphogens), the negative values are accommodated (explained) by considering the solutions as oscillations (departures from, perurbations around) equilibrium (steady state) values ([2], p47).

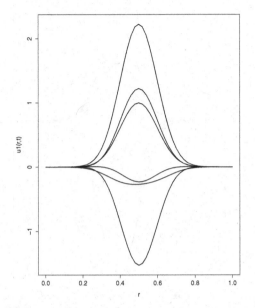

Figure 4.1a: Numerical solution of eq. (4.1a) for $u_1(r,t)$, ncase=4, from `matplot`

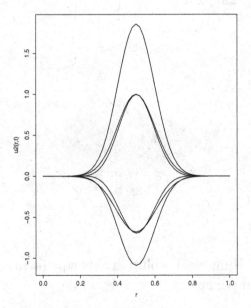

Figure 4.1b: Numerical solution of eq. (4.1b) for $u_2(r,t)$, ncase=4, from `matplot`

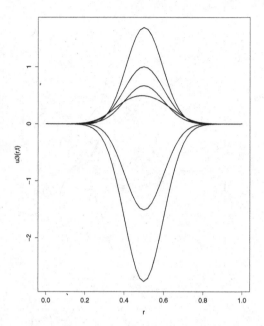

Figure 4.1c: Numerical solution of eq. (4.1c) for $u_3(r,t)$, ncase=4, from `matplot`

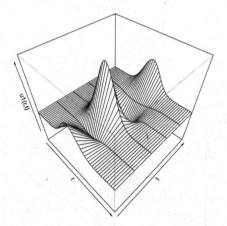

Figure 4.1d: Numerical solution of eq. (4.1a) for $u_1(r,t)$, ncase=4, from `persp`

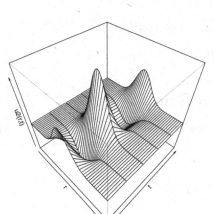

Figure 4.1e: Numerical solution of eq. (4.1b) for $u_2(r,t)$, ncase=4, from `persp`

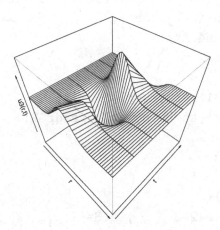

Figure 4.1f: Numerical solution of eq. (4.1c) for $u_3(r,t)$, ncase=4, from `persp`

The complex (figuratively and mathematically) solutions of Figs. 4.1 give an indication of the departure from just diffusion that is possible with the RD system of PDEs (Turing models).

Figs. 4.1 indicate an oscillation in t from diffusion (dispersion) and reaction. Also, n=51 and nout=6 appear to give acceptable resolution in r and t.

(4.2) Eigenvalue Analysis

An eigenvalue analysis of eqs. (4.1) follows.

(4.2.1) Main program

A main program for the eigenvalue analysis of eqs. (4.1) follows.

```
#
# 1D 3 x 3 RD, cylindrical
#
# Delete previous workspaces
  rm(list=ls(all=TRUE))
#
# Access functions for numerical solution
  setwd("f:/turing/jacob");
  source("eigen4.R");
#
# Grid in r
  rl=0;ru=1;n=11;dr=(ru-rl)/(n-1);
  r=seq(from=rl,to=ru,by=dr);drs=dr^2;
#
# Parameters
  ncase=4;
  if(ncase==4){
  a11=-10/3 ; a12=3       ; a13=-1  ;
  a21=-2    ; a22=7/3     ; a23=0   ;
  a31=3     ; a32=-4      ; a33=0   ;
  D1=2/3*drs; D2=1/3*drs; D3=0*drs;}
  if(ncase==5){
  a11=-1  ; a12=-1  ; a13= 0  ;
  a21= 1  ; a22= 0  ; a23=-1  ;
  a31= 0  ; a32= 1  ; a33= 0  ;
  D1=1*drs; D2=0*drs; D3=0*drs;}
#
```

```
# Factors used in eigen4.R
  D1dr2=D1/dr^2   ;D2dr2=D2/dr^2  ;
  D3dr2=D3/dr^2  ;
  D12dr=D1/(2*dr);D22dr=D2/(2*dr);
  D32dr=D3/(2*dr);
#
# Base dependent variable vector
  n2=2*n;n3=3*n;
  ub=rep(0,n3);
# for(i in 1:n3){
#   cat(sprintf("\n i = %2d  ub[i] = %8.4f",
#               i,ub[i]));
# }
#
# Base dependent variable derivative vector
  utb=eigen4(t,ub,parm);
# for(i in 1:n3){
#   cat(sprintf("\n i = %2d  utb[i] = %8.4f",
#               i,utb[i]));
# }
#
# Incremented dependent variable vector
  u=rep(0,n3);ut=rep(0,n3);
  J=matrix(0,nrow=n3,ncol=n3);
#
# Step through Jacobian matrix columns
  for(j in 1:n3){
    u[j]=ub[j]+0.01;
    ut=eigen4(t,u,parm);
#
#   Step through Jacobian matrix rows
    for(i in 1:n3){
      J[i,j]=(ut[i]-utb[i])/(u[j]-ub[j]);
      cat(sprintf("\n i = %2d  j = %2d
```

```
#        J(i,j) = %8.4f",i,j,J[i,j]));
#
#    Next row
     }
     u[j]=ub[j];
#    cat(sprintf("\n"));
#
# Next column
     }
#
# Compute and display eigenvalues
     lam=eigen(J,only.values=TRUE);
     lamVec=lam$values;
     Re_lam=rep(0,n3);Im_lam=rep(0,n3);
     iout=0;
     for(i in 1:n3){
       Re_lam[i]=Re(lamVec[i]);Im_lam[i]=Im(lamVec[i]);
#      if(Re_lam[i]>0){
         cat(sprintf(
           "\n i = %2d   Re = %8.4f   Im = %8.4f",
           i,Re_lam[i],Im_lam[i]));
#        iout=iout+1;
#      }
     }
     cat(sprintf("\n iout = %3d",iout));
#
# Plot eigenvalues
     Re_0=rep(0,n3);Im_0=rep(0,n3);
     Im_0[1]= -1;Im_0[n3]= 1;
#    Im_0[1]=-40;Im_0[n3]=40;
     plot(Re_lam,Im_lam,lwd=2,col="black",pch="o",
          xlab="Re_lam",ylab="Im_lam",main="");
     lines(Re_0,Im_0,type="l",lwd=2);
```

Listing 4.3: Main program for eqs. (4.1)

We can note the following details about Listing 4.3 (similar to Listing 3.3).

- Previous workspaces are deleted. The `setwd` (set working directory) requires editing for the local computer to specify the directory (folder) with the R routines (note the use of / rather than the usual \). `eigen4` is a function for the $3n \times 3n$ linear, MOL system for eqs. (4.1) (discussed next).

```
#
# 1D 3 x 3 RD, cylindrical
#
# Delete previous workspaces
  rm(list=ls(all=TRUE))
#
# Access functions for numerical solution
  setwd("f:/turing/jacob");
  source("eigen4.R");
```

- A uniform grid in r of 11 points is defined with the `seq` utility for the interval $r_l = 0 \leq r \leq r_u = 1$. Therefore, the vector x has the values $r = 0, 1/10, \ldots, 1$.

```
#
# Grid in r
  rl=0;ru=1;n=11;dr=(ru-rl)/(n-1);
  r=seq(from=rl,to=ru,by=dr);drs=dr^2;
```

The small number of grid points, $n = 11$, was selected to keep the output (if selected by deactivating comments) at a more manageable level. This main program also executes with $n = 51$ (as discussed previously in Listing 4.1).

- Two cases are programmed (for the cases in [2], pp 53-54).

```
#
# Parameters
```

```
      ncase=4;
      if(ncase==4){
      a11=-10/3 ; a12=3      ; a13=-1  ;
      a21=-2     ; a22=7/3   ; a23=0    ;
      a31=3      ; a32=-4    ; a33=0    ;
      D1=2/3*drs; D2=1/3*drs; D3=0*drs;}
      if(ncase==5){
      a11=-1  ; a12=-1  ; a13= 0  ;
      a21= 1  ; a22= 0  ; a23=-1  ;
      a31= 0  ; a32= 1  ; a33= 0  ;
      D1=1*drs; D2=0*drs; D3=0*drs;}
```

- Factors used in `eigen4.R` are computed.

```
      #
      # Factors used in eigen4.R
      D1dr2=D1/dr^2  ;D2dr2=D2/dr^2   ;
      D3dr2=D3/dr^2   ;
      D12dr=D1/(2*dr);D22dr=D2/(2*dr);
      D32dr=D3/(2*dr);  .
```

- Base values ub for u_1, u_2, u_3 (eqs. (4.1)) are given zero values with the `rep` utility.

```
      #
      # Base dependent variable vector
      n2=2*n;n3=3*n;
      ub=rep(0,n3);
      # for(i in 1:n3){
      #   cat(sprintf("\n i = %2d  ub[i] = %8.4f",
      #               i,ub[i]));
      # }
```

The output is suppressed because of limited space in the discussion.

- The derivative vector utb at the base values ub is evaluated by a call to eigen4.

```
#
# Base dependent variable derivative vector
  utb=eigen4(t,ub,parm);
# for(i in 1:n3){
#   cat(sprintf("\n i = %2d  utb[i] = %8.4f",
#                   i,utb[i]));
# }
```

The output is suppressed because of limited space in the discussion.

- Vectors for f_i, u_j in eq. (2.5a) are declared (allocated).

```
#
# Incremented dependent variable vector
  u=rep(0,n3);ut=rep(0,n3);
  J=matrix(0,nrow=n3,ncol=n3);
```

A matrix for the $3n \times 3n$ Jacobian matrix **J** in eq. (2.5b) is declared with the matrix utility.

- Within a given column (j), the rows are varied to implement eq. (2.5c). i is the index for the rows of the Jacobian matrix and corresponds to i in eq. (2.5c) varied in the inner for.

  $u_j + \Delta u_j$ in eq. (2.5c) is defined by a change in u_j in the outer for. An absolute increment 0.01 is used (u[j]=ub[j]+0.01).

  The computation of the i,jth element of the Jacobian matrix is J[i,j]=(ut[i]-utb[i])/(u[j]-ub[j]).

eigen4 is used to calculate the corresponding derivative vector ut.

```
#
# Step through Jacobian matrix columns
  for(j in 1:n3){
    u[j]=ub[j]+0.01;
    ut=eigen4(t,u,parm);
#
#   Step through Jacobian matrix rows
    for(i in 1:n3){
      J[i,j]=(ut[i]-utb[i])/(u[j]-ub[j]);
#     cat(sprintf("\n i = %2d   j = %2d
#        J(i,j) = %8.4f",i,j,J[i,j]));
#
#   Next row
    }
    u[j]=ub[j];
#   cat(sprintf("\n"));
#
# Next column
  }
```

The output is suppressed because of limited space in the discussion.

- At this point, all $3n \times 3n$ elements of the approximate Jacobian matrix have been computed. The $3n$ eigenvalues of **J** can now be computed (numerically, without factoring the characteristic polynomial (2.4b)).

```
#
# Compute and display eigenvalues
  lam=eigen(J,only.values=TRUE);
  lamVec=lam$values;
  Re_lam=rep(0,n3);Im_lam=rep(0,n3);
```

```
      iout=0;
      for(i in 1:n3){
        Re_lam[i]=Re(lamVec[i]);
        Im_lam[i]=Im(lamVec[i]);
  #     if(Re_lam[i]>0){
          cat(sprintf(
            "\n i = %2d   Re = %8.4f   Im = %8.4f",
            i,Re_lam[i],Im_lam[i]));
  #       iout=iout+1;
  #     }
      }
      cat(sprintf("\n iout = %3d",iout));
```

The test for unstable eigenvalues is suppressed so that all
of the eigenvalues are displayed.

- The eigenvalues are plotted as points (circles, pch="o") in
  the complex plane. A line is added (lines) for a zero real
  part (the boundary between stable and unstable eigenvalues) using two vectors, Re_0, Im_0.

```
  #
  # Plot eigenvalues
    Re_0=rep(0,n3);Im_0=rep(0,n3);
    Im_0[1]= -1;Im_0[n3]= 1;
  # Im_0[1]=-40;Im_0[n3]=40;
    plot(Re_lam,Im_lam,lwd=2,col="black",pch="o",
         xlab="Re_lam",ylab="Im_lam",main="");
    lines(Re_0,Im_0,type="l",lwd=2);
```

Automatic scaling of the abscissa (horizontal, x) and
ordinate (vertical, y) axes accommodates variations in
the plots for ncase=4,5 (Fig. 4.2 for ncase=4).

The subordinate routine eigen4 called by the main program
of Listing 4.3 is considered next.

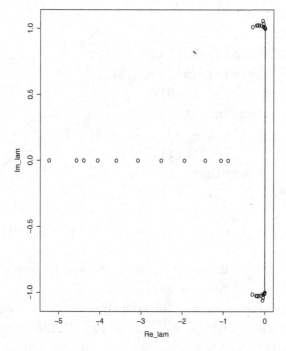

Figure 4.2: Eigenvalues for `ncase=4`

(4.2.2) Subordinate routine

The subordinate routine for the eqs. (4.1) follows.

```
  eigen4=function(t,u,parm){
#
# Function eigen4 computes the t derivative
# vector for u1(r,t), u2(r,t), u3(r,t)
#
# One vector to three vectors
  u1 =rep(0,n);u2 =rep(0,n);u3 =rep(0,n);
  u1t=rep(0,n);u2t=rep(0,n);u3t=rep(0,n);
  for(i in 1:n){
   u1[i]=u[i];
    u2[i]=u[i+n];
```

```
      u3[i]=u[i+n2];
   }
#
# u1t(r,t)
  for(i in 1:n){
    if(i==1){u1t[1]=a11*u1[1]+a12*u2[1]+a13*u3[1]+
             4*D1dr2*(u1[  2]-u1[1]);}
    if(i==n){u1t[n]=a11*u1[n]+a12*u2[n]+a13*u3[n]+
             2*D1dr2*(u1[n-1]-u1[n]);}
    if((i>1)&(i<n)){
       u1t[i]=a11*u1[i]+a12*u2[i]+a13*u3[i]+
             D1dr2*(u1[i+1]-2*u1[i]+u1[i-1])+
             D12dr*(1/r[i])*(u1[i+1]-u1[i-1]);}
  }
#
# u2t(r,t)
  for(i in 1:n){
    if(i==1){u2t[1]=a21*u1[1]+a22*u2[1]+a23*u3[1]+
             4*D2dr2*(u2[  2]-u2[1]);}
    if(i==n){u2t[n]=a21*u1[n]+a22*u2[n]+a23*u3[n]+
             2*D2dr2*(u2[n-1]-u2[n]);}
    if((i>1)&(i<n)){
       u2t[i]=a21*u1[i]+a22*u2[i]+a23*u3[i]+
             D2dr2*(u2[i+1]-2*u2[i]+u2[i-1])+
             D22dr*(1/r[i])*(u2[i+1]-u2[i-1]);}
  }
#
# u3t(r,t)
  for(i in 1:n){
    if(i==1){u3t[1]=a31*u1[1]+a32*u2[1]+a33*u3[1]+
             4*D3dr2*(u3[  2]-u3[1]);}
    if(i==n){u3t[n]=a31*u1[n]+a32*u2[n]+a33*u3[n]+
             2*D3dr2*(u3[n-1]-u3[n]);}
    if((i>1)&(i<n)){
```

```
        u3t[i]=a31*u1[i]+a32*u2[i]+a33*u3[i]+
            D3dr2*(u3[i+1]-2*u3[i]+u3[i-1])+
            D32dr*(1/r[i])*(u3[i+1]-u3[i-1]);}
  }
#
# Three vectors to one vector
  ut=rep(0,n3);
  for(i in 1:n){
    ut[i]    =u1t[i];
    ut[i+n]  =u2t[i];
    ut[i+n2]=u3t[i];
  }
#
# Return derivative vector
  return(c(ut));
  }
```

Listing 4.4: Routine `eigen4` for eqs. (4.1)

`eigen4` is very similar to `pde_1d` of Listing 4.2. Only the differences are noted next.

- The function is defined. `t` is the current value of t in eqs. (4.1) (unused in the eigenvalue analysis). `u` is the $3n = 3(11) = 33$-vector of ODE dependent variables.

  ```
  eigen4=function(t,u,parm){
  #
  # Function eigen4 computes the t derivative
  # vector for u1(r,t), u2(r,t), u3(r,t)
  ```

- The derivative vector is returned to the main program. `c` is the R utility for a numerical vector (a `list` is also used in `pde_1d` of Listing 4.2 as required by `lsodes`).

```
#
# Return derivative vector
  return(c(ut));
  }
```

The final } concludes `eigen4`.

The numerical output from the routines in Listings 4.3, 4.4 is next.

(4.2.3) Model output

For `ncase=4` in Listing 4.3, the output (with the full set of eigenvalues included) is

```
i =  1  Re =  -5.2377  Im =   0.0000
i =  2  Re =  -4.5737  Im =   0.0000
i =  3  Re =  -4.3945  Im =   0.0000
i =  4  Re =  -4.0590  Im =   0.0000
i =  5  Re =  -3.6078  Im =   0.0000
i =  6  Re =  -3.0796  Im =   0.0000
i =  7  Re =  -2.5138  Im =   0.0000
i =  8  Re =  -1.9521  Im =   0.0000
i =  9  Re =  -1.4449  Im =   0.0000
i = 10  Re =  -1.0644  Im =   0.0000
i = 11  Re =  -0.0552  Im =   1.0588
i = 12  Re =  -0.0552  Im =  -1.0588
i = 13  Re =  -0.3021  Im =   1.0120
i = 14  Re =  -0.3021  Im =  -1.0120
i = 15  Re =  -0.2069  Im =   1.0236
i = 16  Re =  -0.2069  Im =  -1.0236
i = 17  Re =  -0.1833  Im =   1.0246
i = 18  Re =  -0.1833  Im =  -1.0246
i = 19  Re =  -0.0405  Im =   1.0366
i = 20  Re =  -0.0405  Im =  -1.0366
i = 21  Re =  -0.1416  Im =   1.0246
```

```
i = 22   Re =   -0.1416   Im =   -1.0246
i = 23   Re =   -0.0918   Im =    1.0210
i = 24   Re =   -0.0918   Im =   -1.0210
i = 25   Re =   -0.0446   Im =    1.0134
i = 26   Re =   -0.0446   Im =   -1.0134
i = 27   Re =   -0.0144   Im =    1.0098
i = 28   Re =   -0.0144   Im =   -1.0098
i = 29   Re =   -0.0110   Im =    1.0043
i = 30   Re =   -0.0110   Im =   -1.0043
i = 31   Re =   -0.0001   Im =    1.0001
i = 32   Re =   -0.0001   Im =   -1.0001
i = 33   Re =   -0.8896   Im =    0.0000

iout =   0
```

Table 4.2: Numerical output for **ncase=4**, Listing 4.3

We can note the following details about this output.

- The real part of all 33 eigenvalues is negative so the PDE system of eqs. (4.1) programmed in Listings 4.1, 4.2 is stable.
- Complex eigenvalue pairs indicate the system of eqs. (4.1) is oscillatory, as demonstrated in Figs. 4.1.
- **iout** = 0 also indicates there are no unstable eigenvalues.

Fig. 4.2 indicates that all of the eigenvalues have negative real parts with some complex conjugate pairs as reflected in the output of Table 4.2.

ncase=5 in Listing 4.1 is not considered in detail to conserve space, but the reader can easily execute this case.

(4.3) 1D 3 × 3 Linear Model in Spherical Coordinates

An analysis in spherical coordinates (r, θ, ϕ) closely parallels the preceding analysis in cylindrical coordinates (r, θ, z). For just the radial variation, the PDEs are

$$\frac{\partial u_1}{\partial t} = a_{11}u_1 + a_{12}u_2 + a_{13}u_3 + D_1 \left(\frac{\partial^2 u_1}{\partial r^2} + \frac{2}{r}\frac{\partial u_1}{\partial r} \right) \quad \text{(4.4a)}$$

$$\frac{\partial u_2}{\partial t} = a_{21}u_1 + a_{22}u_2 + a_{23}u_3 + D_2 \left(\frac{\partial^2 u_2}{\partial r^2} + \frac{2}{r}\frac{\partial u_2}{\partial r} \right) \quad \text{(4.4b)}$$

$$\frac{\partial u_3}{\partial t} = a_{31}u_1 + a_{32}u_2 + a_{43}u_3 + D_3 \left(\frac{\partial^2 u_3}{\partial r^2} + \frac{2}{r}\frac{\partial u_3}{\partial r} \right) \quad \text{(4.4c)}$$

The difference between PDEs (4.1) and (4.4) is in the variable coefficients $\frac{1}{r}$ (cylindrical) and $\frac{2}{r}$ (spherical). This difference may seem inconsequential, but the difference in the geometry between the two cases is important. For example, spherical coordinates might be a good choice to represent a tumor (which could not be very well represented in Cartesian or cylindrical coordinates).

Because the two sets of PDEs are so similar, the difference in the coding in the ODE/MOL routines is easily summarized.

Cylindrical, Listing 4.2

```
u1t

    4*D1dr2*(u1[2]-u1[1])

    D12dr*(1/r[i])*(u1[i+1]-u1[i-1])

u2t

    4*D2dr2*(u2[2]-u2[1])
```

```
        D22dr*(1/r[i])*(u2[i+1]-u2[i-1])

    u3t

        4*D3dr2*(u3[2]-u3[1])

        D32dr*(1/r[i])*(u3[i+1]-u3[i-1])

Spherical

    u1t

        6*D1dr2*(u1[2]-u1[1])

        D12dr*(2/r[i])*(u1[i+1]-u1[i-1])

    u2t

        6*D2dr2*(u2[2]-u2[1])

        D22dr*(2/r[i])*(u2[i+1]-u2[i-1])

    u3t

        6*D3dr2*(u3[2]-u3[1])

        D32dr*(2/r[i])*(u3[i+1]-u3[i-1])
```

As might be expected, the two solutions are similar (this can easily be verified by the reader through a modification of pde_1d and **eigen4** in Listings 4.2 and 4.4). More substantive differences in the coding and solutions would be required if the other two cylindrical (θ, z) and spherical (θ, ϕ) coordinates are included since the 2D and 3D PDEs are substantially different in these coordinates.

(4.4) Summary and Conclusions

Computer-based experimentation to investigate the features of the $1D$ $3n \times 3n$ linear systems in cylindrical and spherical coordinates is straightforward within the MOL framework. This type of investigation can be enhanced by computing and displaying the PDE RHS terms (typically spatial derivatives and source terms) and the LHS terms (usually derivatives in time). In this way, the contributions of the individual PDE terms can be observed and evaluated. This insight may in turn suggest modifications of the model, e.g., addition or removal of terms, additional PDEs, with, ideally, convergence to a useful mathematical model that might, for example, achieve agreement with reported experimental data or observed features/phenomena.

References

[1] Schiesser, W.E. (2013), *Partial Differential Equation Analysis in Biomedical Engineering*, Cambridge University Press, Cambridge, UK

[2] Turing, A.M. (1952), The chemical basis of morphogenesis, *Philosophical Transactions of the Royal Society of London, Series B, Biological Sciences*, **237**, no. 641

Chapter 5

Two Dimensional PDEs

The reaction-diffusion (RD) models considered in the preceding chapters are 1D in the Cartesian variable x or the cylindrical and spherical radial coordinate r. We now consider the extension to 2D in Cartesian coordinates x, y.

(5.1) Linear Reaction-diffusion Model

The 2D RD equations considered in this chapter (an extension of eqs. (1.5)) are

$$\frac{\partial u_1}{\partial t} = a_{11}u_1 + a_{12}u_2 + a_{13}u_3 + D_1\frac{\partial^2 u_1}{\partial x^2} + D_1\frac{\partial^2 u_1}{\partial y^2} \qquad (5.1a)$$

$$\frac{\partial u_2}{\partial t} = a_{21}u_1 + a_{22}u_2 + a_{23}u_3 + D_2\frac{\partial^2 u_2}{\partial x^2} + D_2\frac{\partial^2 u_2}{\partial y^2} \qquad (5.1b)$$

$$\frac{\partial u_3}{\partial t} = a_{31}u_1 + a_{32}u_2 + a_{33}u_3 + D_3\frac{\partial^2 u_3}{\partial x^2} + D_3\frac{\partial^2 u_3}{\partial y^2} \qquad (5.1c)$$

where $a_{11}, \ldots, a_{33}$ are constants to be specified. Eqs. (5.1) are linear, constant coefficient PDEs that can be analyzed by the usual method of lines (MOL) if the diffusion terms are replaced by algebraic approximations such as finite differences (FDs). The resulting system of ODEs in t are the starting point for the MOL analysis that follows. The MOL based on FDs is an extension of the 1D formulation considered in [2] and the MOL/ODEs are again termed a *Turing model*.

171

Eqs. (5.1) are first order in t and second order in x, y. They therefore each require one initial condition (IC) and two boundary conditions (BCs) in x and y.

$$u_1(x, y, t = 0) = f_1(x, y); \ u_2(x, y, t = 0) = f_2(x, y)$$

$$u_3(x, y, t = 0) = f_3(x, y); \qquad (5.2a,b,c)$$

$$\frac{\partial u_1(x = x_l, y, t)}{\partial x} = \frac{\partial u_1(x = x_u, y, t)}{\partial x} = 0 \qquad (5.3a,b)$$

$$\frac{\partial u_2(x = x_l, y, t)}{\partial x} = \frac{\partial u_2(x = x_u, y, t)}{\partial x} = 0 \qquad (5.3c,d)$$

$$\frac{\partial u_3(x = x_l, y, t)}{\partial x} = \frac{\partial u_3(x = x_u, y, t)}{\partial x} = 0 \qquad (5.3e,f)$$

$$\frac{\partial u_1(x, y = y_l, t)}{\partial y} = \frac{\partial u_1(x, y = y_u, t)}{\partial y} = 0 \qquad (5.4a,b)$$

$$\frac{\partial u_2(x, y = y_l, t)}{\partial y} = \frac{\partial u_2(x, y = y_u, t)}{\partial y} = 0 \qquad (5.4c,d)$$

$$\frac{\partial u_3(x, y = y_l, t)}{\partial y} = \frac{\partial u_3(x, y = y_u, t)}{\partial y} = 0 \qquad (5.4e,f)$$

Eqs. (5.2), (5.3) and (5.4) constitute the 2D 3×3 RD PDE system which is analyzed with the following routines.

(5.2) Computer Routines, Finite Differences

The computer analysis of the preceding PDE system is implemented with a series of routines that are next listed and discussed in detail.

(5.2.1) Main program

A main program for eqs. (5.1) follows.

```
#
# 2D 3 x 3 Turing, FDs
```

```
#
# Delete previous workspaces
  rm(list=ls(all=TRUE))
#
# Access ODE integrator
  library("deSolve");
#
# Access functions for numerical solution
  setwd("f:/turing/chap5");
  source("pde_1e.R");
  source("dss042.R");
  source("dss044.R");
  source("dss046.R");
#
# Grid in x,y (same n for x,y)
  xl=0;xu=1;n=21;dx=(xu-xl)/(n-1);
  x=seq(from=xl,to=xu,by=dx);
  yl=0;yu=1;dy=(yu-yl)/(n-1);
  y=seq(from=yl,to=yu,by=dy);
  n2=n^2;n22=2*n2;n3=3*n2;
#
# Parameters
  ncase=4;
  if(ncase==1){
    a11  =0; a12  =0; a13  =0;
    a21  =0; a22  =0; a23  =0;
    a31  =0; a32  =0; a33  =0;
    D1=dx^2; D2=dx^2; D3=dx^2;
    t0   =0; tf   =2; nout =4;
    c1   =0; c2   =50;
    zl   =0; zu   =1;}
  if(ncase==2){
    a11= -1; a12  =0; a13  =0;
    a21 = 0; a22  =-1; a23 = 0;
```

```
   a31 = 0; a32   =0; a33 =-1;
   D1=dx^2; D2=dx^2; D3=dx^2;
   t0   =0; tf   =2; nout =4;
   c1   =0; c2   =50;
   zl   =0; zu   =1;}
 if(ncase==3){
   a11  =1; a12  =0; a13  =0;
   a21  =0; a22  =1; a23  =0;
   a31  =0; a32  =0; a33  =1;
   D1=dx^2; D2=dx^2; D3=dx^2;
   t0   =0; tf   =2; nout =4;
   c1   =0; c2   =50;
   zl   =0; zu   =1;}
 if(ncase==4){
 a11   =-10/3; a12      =3; a13    =-1;
 a21        =-2; a22    =7/3; a23    =0;
 a31        =3; a32     =-4; a33    =0;
 D1=2/3*dx^2; D2=1/3*dx^2; D3=0*dx^2;
 t0         =0; tf        =20; nout   =4;
 c1         =0; c2        =50;
 zl         =-2; zu        =3;}
 if(ncase==5){
 a11   =-1; a12   =-1; a13   = 0;
 a21   =1; a22   =0; a23   =-1;
 a31   =0; a32   =1; a33   =0;
 D1=1*dx^2; D2=0*dx^2; D3=0*dx^2;
 t0   =0; tf   =2; nout   =4;
 c1   =0; c2   =50;
 zl   =-1; zu   =2;}
#
# Independent variable for ODE integration
 tout=seq(from=t0,to=tf,by=(tf-t0)/(nout-1));
#
# ICs
```

```
  u0=rep(0,n3);
  for(i in 1:n){
  for(j in 1:n){
    ij=(i-1)*n+j;
    u0[ij]    =c1+exp(-c2*((x[i]-0.5)^2+
              (y[j]-0.5)^2));
    u0[ij+n2] =c1+exp(-c2*((x[i]-0.5)^2+
              (y[j]-0.5)^2));
    u0[ij+n22]=c1+exp(-c2*((x[i]-0.5)^2+
              (y[j]-0.5)^2));
  }
  }
  ncall=0;
#
# ODE integration
  out=lsodes(y=u0,times=tout,func=pde_1e,
     sparsetype="sparseint",rtol=1e-6,
     atol=1e-6,maxord=5);
  nrow(out)
  ncol(out)
#
# Arrays for numerical solution
#
# it=1
  u1_1=matrix(0,nrow=n,ncol=n);
  u2_1=matrix(0,nrow=n,ncol=n);
  u3_1=matrix(0,nrow=n,ncol=n);
#
# it=2
  u1_2=matrix(0,nrow=n,ncol=n);
  u2_2=matrix(0,nrow=n,ncol=n);
  u3_2=matrix(0,nrow=n,ncol=n);
#
# it=3
```

```
  u1_3=matrix(0,nrow=n,ncol=n);
  u2_3=matrix(0,nrow=n,ncol=n);
  u3_3=matrix(0,nrow=n,ncol=n);
#
# it=4
  u1_4=matrix(0,nrow=n,ncol=n);
  u2_4=matrix(0,nrow=n,ncol=n);
  u3_4=matrix(0,nrow=n,ncol=n);
#
# t variation
  for(it in 1:nout){
#
# x,y variation
  for(i in 1:n){
  for(j in 1:n){
    if(it==1){
      ij=(i-1)*n+j;
      u1_1[i,j]=out[it,ij+1];
      u2_1[i,j]=out[it,ij+1+n2];
      u3_1[i,j]=out[it,ij+1+n22];
    }
    if(it==2){
      ij=(i-1)*n+j;
      u1_2[i,j]=out[it,ij+1];
      u2_2[i,j]=out[it,ij+1+n2];
      u3_2[i,j]=out[it,ij+1+n22];
    }
    if(it==3){
      ij=(i-1)*n+j;
      u1_3[i,j]=out[it,ij+1];
      u2_3[i,j]=out[it,ij+1+n2];
      u3_3[i,j]=out[it,ij+1+n22];
     }
    if(it==4){
```

```
      ij=(i-1)*n+j;
      u1_4[i,j]=out[it,ij+1];
      u2_4[i,j]=out[it,ij+1+n2];
      u3_4[i,j]=out[it,ij+1+n22];
    }
#
# Next y
  }
#
# Next x
  }
  if(it==1){cat(sprintf("\n t0 = %4.2f",
                        tout[it]))};
  if(it==2){cat(sprintf("\n t1 = %4.2f",
                        tout[it]))};
  if(it==3){cat(sprintf("\n t2 = %4.2f",
                        tout[it]))};
  if(it==4){cat(sprintf("\n t3 = %4.2f",
                        tout[it]))};
#
# Next t
  }
  cat(sprintf(" ncall = %4d\n",ncall));
#
# Plot 3D numerical solution
#
# u1 vs t
  par(mfrow=c(2,2))
  persp(x,y,u1_1,theta=45,phi=45,
    xlim=c(xl,xu),ylim=c(yl,yu),zlim=c(zl,zu),
    xlab="x",ylab="y",zlab="u1(x,y,t0)");
  persp(x,y,u1_2,theta=45,phi=45,
    xlim=c(xl,xu),ylim=c(yl,yu),zlim=c(zl,zu),
    xlab="x",ylab="y",zlab="u1(x,y,t1)");
```

```
  persp(x,y,u1_3,theta=45,phi=45,
    xlim=c(xl,xu),ylim=c(yl,yu),zlim=c(zl,zu),
    xlab="x",ylab="y",zlab="u1(x,y,t2)");
  persp(x,y,u1_4,theta=45,phi=45,
    xlim=c(xl,xu),ylim=c(yl,yu),zlim=c(zl,zu),
    xlab="x",ylab="y",zlab="u1(x,y,t3)");
#
# u2 vs t
  par(mfrow=c(2,2))
  persp(x,y,u2_1,theta=45,phi=45,
    xlim=c(xl,xu),ylim=c(yl,yu),zlim=c(zl,zu),
    xlab="x",ylab="y",zlab="u2(x,y,t0)");
  persp(x,y,u2_2,theta=45,phi=45,
    xlim=c(xl,xu),ylim=c(yl,yu),zlim=c(zl,zu),
    xlab="x",ylab="y",zlab="u2(x,y,t1)");
  persp(x,y,u2_3,theta=45,phi=45,
    xlim=c(xl,xu),ylim=c(yl,yu),zlim=c(zl,zu),
    xlab="x",ylab="y",zlab="u2(x,y,t2)");
  persp(x,y,u2_4,theta=45,phi=45,
    xlim=c(xl,xu),ylim=c(yl,yu),zlim=c(zl,zu),
    xlab="x",ylab="y",zlab="u2(x,y,t3)");
#
# u3 vs t
  par(mfrow=c(2,2))
  persp(x,y,u3_1,theta=45,phi=45,
    xlim=c(xl,xu),ylim=c(yl,yu),zlim=c(zl,zu),
    xlab="x",ylab="y",zlab="u3(x,y,t0)");
  persp(x,y,u3_2,theta=45,phi=45,
    xlim=c(xl,xu),ylim=c(yl,yu),zlim=c(zl,zu),
    xlab="x",ylab="y",zlab="u3(x,y,t1)");
  persp(x,y,u3_3,theta=45,phi=45,
    xlim=c(xl,xu),ylim=c(yl,yu),zlim=c(zl,zu),
    xlab="x",ylab="y",zlab="u3(x,y,t2)");
  persp(x,y,u3_4,theta=45,phi=45,
```

```
xlim=c(xl,xu),ylim=c(yl,yu),zlim=c(zl,zu),
xlab="x",ylab="y",zlab="u3(x,y,t3)");
```

Listing 5.1: Main program for eqs. (5.1), finite differences

We can note the following details about Listing 5.1.

- Previous workspaces are removed. Then the ODE integrator library deSolve is accessed. Note that the setwd
(set working directory) uses / rather than the usual \.

```
#
# 2D 3 x 3 Turing, FDs
#
# Delete previous workspaces
  rm(list=ls(all=TRUE))
#
# Access ODE integrator
  library("deSolve");
#
# Access functions for numerical solution
  setwd("f:/turing/chap5");
  source("pde_1e.R");
  source("dss042.R");
  source("dss044.R");
  source("dss046.R");
```

pde_1e is the routine for the MOL ODEs (discussed subsequently). dss042,dss044,dss046 are library spatial
differentiation routines (DSS = Differentiation in Space
Subroutine).

- Uniform grids of 21 points in x, y are defined with the
seq utility for the interval $x_l = 0 \leq x \leq x_u = 1$, $y_l = 0 \leq y \leq y_u = 1$. Therefore, the vectors x,y have the values
$x = 0, 0.05, \ldots, 1$, $y = 0, 0.05, \ldots, 1$.

```
#
# Grid in x,y (same n for x,y)
  xl=0;xu=1;n=21;dx=(xu-xl)/(n-1);
  x=seq(from=xl,to=xu,by=dx);
  yl=0;yu=1;dy=(yu-yl)/(n-1);
  y=seq(from=yl,to=yu,by=dy);
  n2=n^2;n22=2*n2;n3=3*n2;
```

- Five cases are programmed with variations in the model parameters

```
#
# Parameters
  ncase=4;
  if(ncase==1){
    a11  =0; a12  =0; a13  =0;
    a21  =0; a22  =0; a23  =0;
    a31  =0; a32  =0; a33  =0;
    D1=dx^2; D2=dx^2; D3=dx^2;
    t0   =0; tf   =2; nout =4;
    c1   =0; c2   =50;
    zl   =0; zu   =1;}
        .        .
        .
        .        .
  if(ncase==5){
  a11   =-1; a12   =-1; a13   = 0;
  a21   =1; a22   =0; a23   =-1;
  a31   =0; a32   =1; a33   =0;
  D1=1*dx^2; D2=0*dx^2; D3=0*dx^2;
  t0    =0; tf    =2; nout  =4;
  c1    =0; c2    =50;
  zl    =-1; zu    =2;}
```

These cases are discussed in Chapter 1 after Listing 1.1.

Each diffusivity has a multiplication by dx^2 which cancels the same factor in the denominator of the FD approximations (with $dy^2 = dx^2$). This use of D1,D2,D3 is then consistent with the FD/MOL approximations in [2].

- A uniform grid in t of 4 output points is defined with the seq utility for the interval $t_0 = 0 \le t \le t_f = 2$ or 20 (depending on the case). Therefore, the vector tout has the values $t = 0, 2/3, \ldots, 2$ or $t = 0, 20/3, \ldots, 20$ (which are confirmed in the subsequent output).

```
#
# Independent variable for ODE integration
  tout=seq(from=t0,to=tf,by=(tf-t0)/(nout-1));
```

At this point, the intervals of x, y and t in eqs. (5.1) are defined.

- IC functions in eqs. (5.2a,b,c) are defined. n2,n22,n3 are constants defined previously.

```
#
# ICs
  u0=rep(0,n3);
  for(i in 1:n){
  for(j in 1:n){
    ij=(i-1)*n+j;
    u0[ij]    =c1+exp(-c2*((x[i]-0.5)^2+
              (y[j]-0.5)^2));
    u0[ij+n2] =c1+exp(-c2*((x[i]-0.5)^2+
              (y[j]-0.5)^2));
    u0[ij+n22]=c1+exp(-c2*((x[i]-0.5)^2+
              (y[j]-0.5)^2));
  }
  }
  ncall=0;
```

Two **fors** are used for x and y with subscripts i and j, respectively. These subscripts are combined into a single subscript ij=(i-1)*n+j when a 1D array (vector) is used for the MOL analysis. In other words, ij=(i-1)*n+j is a general approach in going between 1D and 2D arrays. The counter for the calls to **pde_1e** is initialized (and passed to **pde_1e** without a special designation).

- The system of $3(21^2) = 1323$ MOL/ODEs is integrated by the library integrator **lsodes** (available in **deSolve**) with the sparse matrix option specified. As expected, the inputs to **lsodes** are the ODE function, **pde_1e**, the IC vector u0, and the vector of output values of t, **tout**. The length of u0 (e.g., 1323) informs **lsodes** how many ODEs are to be integrated. **func,y,times** are reserved names.

```
#
# ODE integration
  out=lsodes(y=u0,times=tout,func=pde_1e,
     sparsetype="sparseint",rtol=1e-6,
     atol=1e-6,maxord=5);
  nrow(out)
  ncol(out)
```

The numerical solution to the ODEs is returned in matrix **out**. In this case, **out** has the dimensions $nout \times (3n^2 + 1)$ = 4×1324. The offset $(3n^2 + 1)$ is required since the first element of each column has the output t (also in **tout**), and the $2, \ldots, 3n^2 + 1 = 2, \ldots, 1324$ column elements have the 1324 ODE solutions. This indexing is used next.

- Arrays are defined (preallocated) for u_1, u_2, u_3 at the $nout = 4$ values of t.

```
#
# Arrays for numerical solution
```

```
#
# it=1
  u1_1=matrix(0,nrow=n,ncol=n);
  u2_1=matrix(0,nrow=n,ncol=n);
  u3_1=matrix(0,nrow=n,ncol=n);
#
# it=2
  u1_2=matrix(0,nrow=n,ncol=n);
  u2_2=matrix(0,nrow=n,ncol=n);
  u3_2=matrix(0,nrow=n,ncol=n);
#
# it=3
  u1_3=matrix(0,nrow=n,ncol=n);
  u2_3=matrix(0,nrow=n,ncol=n);
  u3_3=matrix(0,nrow=n,ncol=n);
#
# it=4
  u1_4=matrix(0,nrow=n,ncol=n);
  u2_4=matrix(0,nrow=n,ncol=n);
  u3_4=matrix(0,nrow=n,ncol=n);
```

- The ODE solution is placed in a set of $3 \times 4 = 12$ matrices, u1_1 to u3_4, for subsequent plotting by stepping through the solution with respect to x, y and t (within three fors with indices it,i,j for t, x, y, respectively).

```
#
# t variation
  for(it in 1:nout){
#
# x,y variation
  for(i in 1:n){
  for(j in 1:n){
    if(it==1){
      ij=(i-1)*n+j;
```

```
        u1_1[i,j]=out[it,ij+1];
        u2_1[i,j]=out[it,ij+1+n2];
        u3_1[i,j]=out[it,ij+1+n22];
      }
      if(it==2){
        ij=(i-1)*n+j;
        u1_2[i,j]=out[it,ij+1];
        u2_2[i,j]=out[it,ij+1+n2];
        u3_2[i,j]=out[it,ij+1+n22];
      }
      if(it==3){
        ij=(i-1)*n+j;
        u1_3[i,j]=out[it,ij+1];
        u2_3[i,j]=out[it,ij+1+n2];
        u3_3[i,j]=out[it,ij+1+n22];
       }
      if(it==4){
        ij=(i-1)*n+j;
        u1_4[i,j]=out[it,ij+1];
        u2_4[i,j]=out[it,ij+1+n2];
        u3_4[i,j]=out[it,ij+1+n22];
      }
#
# Next y
  }
#
# Next x
  }
  if(it==1){cat(sprintf("\n t0 = %4.2f",
                    tout[it]))};
  if(it==2){cat(sprintf("\n t1 = %4.2f",
                    tout[it]))};
  if(it==3){cat(sprintf("\n t2 = %4.2f",
                    tout[it]))};
```

```
  if(it==4){cat(sprintf("\n t3 = %4.2f",
                        tout[it]))};
#
# Next t
  }
  cat(sprintf(" ncall = %4d\n",ncall));
```

The four output values of t, t0,t1,t2,t3, are displayed
to indicate the progress of the solution.

Finally, the number of calls to pde_1e is displayed to
give an indication of the computational effort required to
compute the solution.

- The solutions to eqs. (5.1), $u_1(x, y, t), u_2(x, y, t), u_3(x, y, t)$, are plotted vs x, y with t as a parameter in 3D
  perspective with **persp** For $u_1(x, y, t)$, four calls to **persp**
  gives four 3D plots on one page with **par(mfrow=c(2,2))**
  (a 2×2 array of plots).

```
#
# Plot 3D numerical solution
#
# u1 vs t
  par(mfrow=c(2,2))
  persp(x,y,u1_1,theta=45,phi=45,
    xlim=c(xl,xu),ylim=c(yl,yu),zlim=c(zl,zu),
    xlab="x",ylab="y",zlab="u1(x,y,t0)");
  persp(x,y,u1_2,theta=45,phi=45,
    xlim=c(xl,xu),ylim=c(yl,yu),zlim=c(zl,zu),
    xlab="x",ylab="y",zlab="u1(x,y,t1)");
  persp(x,y,u1_3,theta=45,phi=45,
    xlim=c(xl,xu),ylim=c(yl,yu),zlim=c(zl,zu),
    xlab="x",ylab="y",zlab="u1(x,y,t2)");
  persp(x,y,u1_4,theta=45,phi=45,
    xlim=c(xl,xu),ylim=c(yl,yu),zlim=c(zl,zu),
    xlab="x",ylab="y",zlab="u1(x,y,t3)");
```

Note that the numbers of rows of **x,y** (n=21) equals the rows and columns of **u1** (**nrows=ncols=n=21**). Similar coding is used for the 3D plotting of $u_2(x, y, t)$, $u_3(x, y, t)$. Scaling in **z** is used with **zlim=c(zl,zu)** so that all four plots in a page can be compared.

The ODE/MOL routine called in the main program of Listing 5.1, **pde_1e**, follows.

(5.2.2) ODE/MOL routine

The ODE/MOL routine for eqs. (5.1) is listed next.

```
  pde_1e=function(t,u,parm){
#
# Function pde_1e computes the t derivative
# vector for u1(x,y,t), u2(x,y,t), u3(x,y,t)
#
# One vector to three arrays
  u1=matrix(0,nrow=n,ncol=n);
  u2=matrix(0,nrow=n,ncol=n);
  u3=matrix(0,nrow=n,ncol=n);
  u1t=matrix(0,nrow=n,ncol=n);
  u2t=matrix(0,nrow=n,ncol=n);
  u3t=matrix(0,nrow=n,ncol=n);
  for(i in 1:n){
  for(j in 1:n){
    ij=(i-1)*n+j;
    u1[i,j]=u[ij];
    u2[i,j]=u[ij+n2];
    u3[i,j]=u[ij+n22];
}
}
#
```

```
# u1x,u2x,u3x,u1y,u2y,u3y
  u1x=matrix(0,nrow=n,ncol=n);
  u2x=matrix(0,nrow=n,ncol=n);
  u3x=matrix(0,nrow=n,ncol=n);
  u1y=matrix(0,nrow=n,ncol=n);
  u2y=matrix(0,nrow=n,ncol=n);
  u3y=matrix(0,nrow=n,ncol=n);
#
# BCs
  nl=2;nu=2;
  for(j in 1:n){
    u1x[1,j]=0;u1x[n,j]=0;
    u2x[1,j]=0;u2x[n,j]=0;
    u3x[1,j]=0;u3x[n,j]=0;
  }
  for(i in 1:n){
    u1y[i,1]=0;u1y[i,n]=0;
    u2y[i,1]=0;u2y[i,n]=0;
    u3y[i,1]=0;u3y[i,n]=0;
  }
#
# u1xx,u2xx,u3xx,u1yy,u2yy,u3yy
  u1xx=matrix(0,nrow=n,ncol=n);
  u2xx=matrix(0,nrow=n,ncol=n);
  u3xx=matrix(0,nrow=n,ncol=n);
  u1yy=matrix(0,nrow=n,ncol=n);
  u2yy=matrix(0,nrow=n,ncol=n);
  u3yy=matrix(0,nrow=n,ncol=n);
  for (j in 1:n){
    u1xx[,j]=dss042(xl,xu,n,u1[,j],u1x[,j],
                    nl,nu);
    u2xx[,j]=dss042(xl,xu,n,u2[,j],u2x[,j],
                    nl,nu);
    u3xx[,j]=dss042(xl,xu,n,u3[,j],u3x[,j],
```

```
                    n1,nu);
  }
  for (i in 1:n){
    u1yy[i,]=dss042(yl,yu,n,u1[i,],u1y[i,],
                    n1,nu);
    u2yy[i,]=dss042(yl,yu,n,u2[i,],u2y[i,],
                    n1,nu);
    u3yy[i,]=dss042(yl,yu,n,u3[i,],u3y[i,],
                    n1,nu);
  }
#
# u1t(x,y,t),u2t(x,y,t),u2t(x,y,t)
  for(i in 1:n){
  for(j in 1:n){
    u1t[i,j]=
      a11*u1[i,j]+a12*u2[i,j]+a13*u3[i,j]+
      D1*u1xx[i,j]+D1*u1yy[i,j];
    u2t[i,j]=
      a21*u1[i,j]+a22*u2[i,j]+a23*u3[i,j]+
      D2*u2xx[i,j]+D2*u2yy[i,j];
    u3t[i,j]=
      a31*u1[i,j]+a32*u2[i,j]+a33*u3[i,j]+
      D3*u3xx[i,j]+D3*u3yy[i,j];
#
# Next j
  }
#
# Next i
  }
#
# Three arrays to one vector
  ut=rep(0,n3);
  for(i in 1:n){
  for(j in 1:n){
```

```
  ij=(i-1)*n+j;
  ut[ij]     =u1t[i,j];
  ut[ij+n2] =u2t[i,j];
  ut[ij+n22]=u3t[i,j];
  }
  }
#
# Increment calls to pde_1e
  ncall <<- ncall+1;
#
# Return derivative vector
  return(list(c(ut)));
  }
```

Listing 5.2: ODE/MOL routine pde_1e for eqs. (5.1), FDs

We can note the following details about pde_1e.

- The function is defined.

```
  pde_1e=function(t,u,parm){
#
# Function pde_1e computes the t derivative
# vector for u1(x,y,t), u2(x,y,t), u3(x,y,t)
```

  t is the current value of t in eqs. (5.1). u is the $3(21^2) =$
  1323-vector of ODE/MOL dependent variables. parm is
  an argument to pass parameters to pde_1e (unused, but
  required in the argument list). The arguments must
  be listed in the order stated to properly interface with
  lsodes called in the main program of Listing 5.1. The
  composite derivative vector of the LHSs of eqs. (5.1) is
  calculated next and returned to lsodes.

- The dependent variable vector is placed in three 2D
  arrays, u1,u2,u3, to facilitate the programming of
  eqs. (5.1).

```
#
# One vector to three arrays
  u1=matrix(0,nrow=n,ncol=n);
  u2=matrix(0,nrow=n,ncol=n);
  u3=matrix(0,nrow=n,ncol=n);
  u1t=matrix(0,nrow=n,ncol=n);
  u2t=matrix(0,nrow=n,ncol=n);
  u3t=matrix(0,nrow=n,ncol=n);
  for(i in 1:n){
  for(j in 1:n){
    ij=(i-1)*n+j;
    u1[i,j]=u[ij];
    u2[i,j]=u[ij+n2];
    u3[i,j]=u[ij+n22];
  }
  }
```

Three arrays are also defined for the LHS derivatives in t of eqs. (5.1), u1t, u2t, u3t. The 1D-2D relation for the indices, ij=(i-1)*n+j, is used as in Listing 5.1.

- Arrays are defined for the first derivatives in x and y, $\dfrac{\partial u_1}{\partial x}$ to $\dfrac{\partial u_3}{\partial y}$.

```
#
# u1x,u2x,u3x,u1y,u2y,u3y
  u1x=matrix(0,nrow=n,ncol=n);
  u2x=matrix(0,nrow=n,ncol=n);
  u3x=matrix(0,nrow=n,ncol=n);
  u1y=matrix(0,nrow=n,ncol=n);
  u2y=matrix(0,nrow=n,ncol=n);
  u3y=matrix(0,nrow=n,ncol=n);
```

- Homogeneous Neumann BCs (5.3), (5.4) are applied.

```
#
# BCs
  nl=2;nu=2;
  for(j in 1:n){
    u1x[1,j]=0;u1x[n,j]=0;
    u2x[1,j]=0;u2x[n,j]=0;
    u3x[1,j]=0;u3x[n,j]=0;
  }
  for(i in 1:n){
    u1y[i,1]=0;u1y[i,n]=0;
    u2y[i,1]=0;u2y[i,n]=0;
    u3y[i,1]=0;u3y[i,n]=0;
  }
```

nl=nu=2 specifies Neumann BCs in the subsequent spatial differentiation (nl=nu=1 would specify Dirichlet BCs).

- Arrays for the second spatial derivatives $\dfrac{\partial^2 u_1}{\partial x^2}$ to $\dfrac{\partial^2 u_3}{\partial y^2}$ are defined.

```
#
# u1xx,u2xx,u3xx,u1yy,u2yy,u3yy
  u1xx=matrix(0,nrow=n,ncol=n);
  u2xx=matrix(0,nrow=n,ncol=n);
  u3xx=matrix(0,nrow=n,ncol=n);
  u1yy=matrix(0,nrow=n,ncol=n);
  u2yy=matrix(0,nrow=n,ncol=n);
  u3yy=matrix(0,nrow=n,ncol=n);
```

- The derivatives $\dfrac{\partial^2 u_1}{\partial x^2}, \dfrac{\partial^2 u_2}{\partial x^2}, \dfrac{\partial^2 u_3}{\partial x^2}$ are computed with the library differentiation routine dss042 for each value of y (from the for).

```
for (j in 1:n){
  u1xx[,j]=dss042(xl,xu,n,u1[,j],u1x[,j],
                  nl,nu);
  u2xx[,j]=dss042(xl,xu,n,u2[,j],u2x[,j],
                  nl,nu);
  u3xx[,j]=dss042(xl,xu,n,u3[,j],u3x[,j],
                  nl,nu);
}
```

Note the use of [,] to specify a range of indices in x. dss042 has three point, second order FDs [1], and is used in place of the explicit FDs discussed in Chapter 1 to simplify the code (the use of the explicit FDs in 2D gives a complicated ODE/MOL routine; consider Listing 1.2 in 2D rather than 1D).

- Similarly, the derivatives $\dfrac{\partial^2 u_1}{\partial y^2}, \dfrac{\partial^2 u_2}{\partial y^2}, \dfrac{\partial^2 u_3}{\partial y^2}$ are computed with dss042 for each value of x (from the for).

```
for (i in 1:n){
  u1yy[i,]=dss042(yl,yu,n,u1[i,],u1y[i,],
                  nl,nu);
  u2yy[i,]=dss042(yl,yu,n,u2[i,],u2y[i,],
                  nl,nu);
```

```
  u3yy[i,]=dss042(yl,yu,n,u3[i,],u3y[i,],
                  nl,nu);
  }
```

Again [,] is used for the range of indices in y.

- The MOL/ODEs of eqs. (5.1) are programmed over the full interval in x and y. The result is the set of ODE derivatives approximating $\dfrac{\partial u_1}{\partial t}, \dfrac{\partial u_2}{\partial t}, \dfrac{\partial u_3}{\partial t}$.

```
#
# u1t(x,y,t),u2t(x,y,t),u2t(x,y,t)
  for(i in 1:n){
  for(j in 1:n){
    u1t[i,j]=
      a11*u1[i,j]+a12*u2[i,j]+a13*u3[i,j]+
      D1*u1xx[i,j]+D1*u1yy[i,j];
    u2t[i,j]=
      a21*u1[i,j]+a22*u2[i,j]+a23*u3[i,j]+
      D2*u2xx[i,j]+D2*u2yy[i,j];
    u3t[i,j]=
      a31*u1[i,j]+a32*u2[i,j]+a33*u3[i,j]+
      D3*u3xx[i,j]+D3*u3yy[i,j];
#
# Next j
  }
#
# Next i
  }
```

- The three arrays with the derivatives in t are placed in one vector, ut (length n3 $= 3(21^2) = 1323$ set in the main program of Listing 5.1).

```
#
# Three arrays to one vector
  ut=rep(0,n3);
  for(i in 1:n){
  for(j in 1:n){
    ij=(i-1)*n+j;
    ut[ij]     =u1t[i,j];
    ut[ij+n2] =u2t[i,j];
    ut[ij+n22]=u3t[i,j];
  }
  }
```

- The number of calls to pde_1e is incremented.

```
#
# Increment calls to pde_1e
  ncall <<- ncall+1;
```

- The derivative vector ut is returned as a list as required by lsodes.

```
#
# Return derivative vector
  return(list(c(ut)));
  }
```

The final } concludes pde_1e.

The numerical and graphical (plotted) output is considered next.

(5.2.3) Model output

For ncase=4, the output from the main program and ODE/MOL routine of Listings 5.1, 5.2 follows.

```
[1] 4

[1] 1324

t0 = 0.00
t1 = 6.67
t2 = 13.33
t3 = 20.00

ncall = 1531
```

Table 5.1: Numerical output for ncase=4

We can note the following details about this output.

- The solution array out from lsodes has the dimensions 4×1324, corresponding to $nout = 4$ and $3(21^2) + 1 = 1323 + 1$ with an offset +1 for the value of t as the first column element at each solution output point. Therefore, the ODE solution vector is in the column elements $2, \ldots, 1323 + 1$.
- The solution output points (values of t) correspond to t0=0, tf=20 as defined in Listing 5.1.

```
t0 = 0.00
t1 = 6.67
t2 = 13.33
t3 = 20.00
```

- The total calls to pde_1e is acceptable, ncall = 1531.

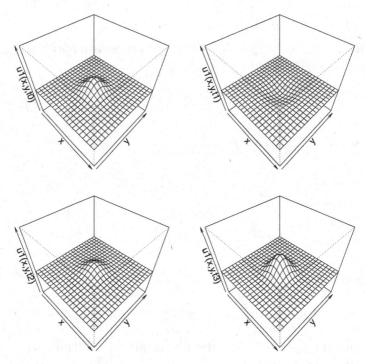

Figure 5.1a: Numerical solution of eq. (5.1a), $u_1(x, y, t)$, FDs, ncase=4

The graphical outout is in Figs. 5.1a ($u_1(x, y, t = 0, 6.67, 13.33, 20)$), 5.1b ($u_2(x, y, t = 0, 6.67, 13.33, 20)$), 5.1c ($u_3(x, y, t = 0, 6.67, 13.33, 20)$).

We can note the following details about Figs. 5.1.

- The ICs for u_1, u_2, u_3 are the same Gaussian function as programmed in Listing 5.1.
- $u_1(x, y, t)$ oscillates in t (through t=t0,t1,t2,t3 for the four plots), in contrast with the case of diffusion only (ncase=1 in Listing 5.1) (this is an example of *Turing oscillation*).
- Similarly, $u_2(x, y, t)$, $u_3(x, y, t)$ oscillate.

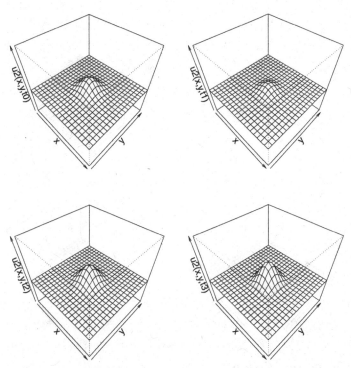

Figure 5.1b: Numerical solution of eq. (5.1b), $u_2(x, y, t)$, FDs, `ncase=4`

Similar results are produced with `ncase=5` in Listing 5.1, that is, the PDE solutions of eqs. (5.1) oscillate rather than a monotonic decay (dispersion) from diffusion only.

In summary, eqs. (5.1) demonstrate oscillation for `ncase=4` with the parameters orginally proposed by Turing. The oscillation could also be displayed with contour plots (rather than the 3D plots from `persp`). This alternative display of the solutions is left as an exercise for the reader.

If the preceding analysis is repeated with `dss044` used in place of `dss042` in `pde_1e` of Listing 5.2, essentially the same results are produced. `dss044` is based on five point, fourth order FDs. This agreement of the solutions for `dss042` and `dss044` demonstrates spatial convergence with $n = 21$ points in x and y.

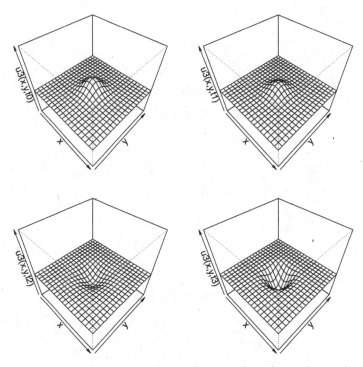

Figure 5.1c: Numerical solution of eq. (5.1c), $u_3(x, y, t)$, FDs, `ncase=4`

A similar conclusion follows from `dss046` in place of `dss042`. `dss046` is based on seven point, sixth order FDs. The ease of interchanging `dss042,dss044,dss046` facilitates this type of *p refinement* (only a name change of the routine is required in `pde_1e` of Listing 5.2).

The preceding analysis is now repeated with the use of splines in place of FDs.

(5.3) Computer Routines, Splines

The main program is not listed since it is similar to the main program of Listing 5.1. The ODE/MOL routine `pde_1f` is discussed in detail subsequently.

(5.3.1) Main program

The only differences between the main programs for FDs (Listing 5.1) and splines are the following.

- ODE/MOL `pde_1f` is accessed in place of `pde_1e`.

```
#
# Access functions for numerical solution
  setwd("f:/turing/chap5");
  source("pde_1f.R");
```

The spline R utility `splinefun` is part of the basic R system and does not have to be accessed separately.
- `pde_1f` is called by `lsodes`.

```
#
# ODE integration
  out=lsodes(y=u0,times=tout,func=pde_1f,
      sparsetype="sparseint",rtol=1e-6,
      atol=1e-6,maxord=5);
  nrow(out)
  ncol(out)
```

Otherwise, the main programs are the same.

(5.3.2) ODE/MOL routine

```
  pde_1f=function(t,u,parm){
#
# Function pde_1f computes the t derivative
# vector for u1(x,y,t), u2(x,y,t), u3(x,y,t)
#
# One vector to three arrays
  u1=matrix(0,nrow=n,ncol=n);
  u2=matrix(0,nrow=n,ncol=n);
  u3=matrix(0,nrow=n,ncol=n);
  u1t=matrix(0,nrow=n,ncol=n);
```

```
u2t=matrix(0,nrow=n,ncol=n);
u3t=matrix(0,nrow=n,ncol=n);
for(i in 1:n){
for(j in 1:n){
  ij=(i-1)*n+j;
  u1[i,j]=u[ij];
  u2[i,j]=u[ij+n2];
  u3[i,j]=u[ij+n22];
}
}
#
# u1x,u2x,u3x
u1x=matrix(0,nrow=n,ncol=n);
u2x=matrix(0,nrow=n,ncol=n);
u3x=matrix(0,nrow=n,ncol=n);
for(j in 1:n){
#
#   u1x
table=splinefun(x,u1[,j]);
u1x[,j]=table(x,deriv=1);
#
#   u2x
table=splinefun(x,u2[,j]);
u2x[,j]=table(x,deriv=1);
#
#   u3x
table=splinefun(x,u3[,j]);
u3x[,j]=table(x,deriv=1);
}
#
# u1y,u2y,u3y
u1y=matrix(0,nrow=n,ncol=n);
u2y=matrix(0,nrow=n,ncol=n);
u3y=matrix(0,nrow=n,ncol=n);
```

```
    for(i in 1:n){
#
#     u1y
      table=splinefun(y,u1[i,]);
      u1y[i,]=table(y,deriv=1);
#
#     u2y
      table=splinefun(y,u2[i,]);
      u2y[i,]=table(y,deriv=1);
#
#     u3y
      table=splinefun(y,u3[i,]);
      u3y[i,]=table(y,deriv=1);
    }
#
# BCs
    for(j in 1:n){
      u1x[1,j]=0;u1x[n,j]=0;
      u2x[1,j]=0;u2x[n,j]=0;
      u3x[1,j]=0;u3x[n,j]=0;
    }
    for(i in 1:n){
      u1y[i,1]=0;u1y[i,n]=0;
      u2y[i,1]=0;u2y[i,n]=0;
      u3y[i,1]=0;u3y[i,n]=0;
    }
#
# u1xx,u2xx,u3xx
    u1xx=matrix(0,nrow=n,ncol=n);
    u2xx=matrix(0,nrow=n,ncol=n);
    u3xx=matrix(0,nrow=n,ncol=n);
    for(j in 1:n){
#
#     u1xx
```

```
      table=splinefun(x,u1x[,j]);
      u1xx[,j]=table(x,deriv=1);
#
#    u2xx
      table=splinefun(x,u2x[,j]);
      u2xx[,j]=table(x,deriv=1);
#
#    u3xx
      table=splinefun(x,u3x[,j]);
      u3xx[,j]=table(x,deriv=1);
   }
#
# u1yy,u2yy,u3yy
   u1yy=matrix(0,nrow=n,ncol=n);
   u2yy=matrix(0,nrow=n,ncol=n);
   u3yy=matrix(0,nrow=n,ncol=n);
   for(i in 1:n){
#
#    u1yy
      table=splinefun(y,u1y[i,]);
      u1yy[i,]=table(y,deriv=1);
#
#    u2yy
      table=splinefun(y,u2y[i,]);
      u2yy[i,]=table(y,deriv=1);
#
#    u3yy
      table=splinefun(y,u3y[i,]);
      u3yy[i,]=table(y,deriv=1);
   }
#
# u1t(x,y,t)
   for(i in 1:n){
   for(j in 1:n){
```

```
    u1t[i,j]=
      a11*u1[i,j]+a12*u2[i,j]+a13*u3[i,j]+
      D1*u1xx[i,j]+D1*u1yy[i,j];
    u2t[i,j]=
      a21*u1[i,j]+a22*u2[i,j]+a23*u3[i,j]+
      D2*u2xx[i,j]+D2*u2yy[i,j];
    u3t[i,j]=
      a31*u1[i,j]+a32*u2[i,j]+a33*u3[i,j]+
      D3*u3xx[i,j]+D3*u3yy[i,j];
#
# Next j
    }
#
# Next i
    }
#
# Three arrays to one vector
  ut=rep(0,n3);
  for(i in 1:n){
  for(j in 1:n){
    ij=(i-1)*n+j;
    ut[ij]     =u1t[i,j];
    ut[ij+n2]  =u2t[i,j];
    ut[ij+n22] =u3t[i,j];
  }
  }
#
# Increment calls to pde_1f
  ncall <<- ncall+1;
#
# Return derivative vector
  return(list(c(ut)));
  }
```

Listing 5.3: ODE/MOL routine `pde_1f` for eqs. (5.1), splines

We can note the following details about pde_1f.

- The function is defined.

```
pde_1f=function(t,u,parm){
#
# Function pde_1f computes the t derivative
# vector for u1(x,y,t), u2(x,y,t), u3(x,y,t)
```

As with pde_1e, t is the current value of t in eqs. (5.1). u is the $3(21^2) = 1323$-vector of ODE/MOL dependent variables. parm is an argument to pass parameters to pde_1e (unused, but required in the argument list). The arguments must be listed in the order stated to properly interface with lsodes called in the main program. The composite derivative vector of the LHSs of eqs. (5.1) is calculated next and returned to lsodes.

- The dependent variable vector is placed in three 2D arrays, u1,u2,u3, to facilitate the programming of eqs. (5.1).

```
#
# One vector to three arrays
  u1=matrix(0,nrow=n,ncol=n);
  u2=matrix(0,nrow=n,ncol=n);
  u3=matrix(0,nrow=n,ncol=n);
  u1t=matrix(0,nrow=n,ncol=n);
  u2t=matrix(0,nrow=n,ncol=n);
  u3t=matrix(0,nrow=n,ncol=n);
  for(i in 1:n){
  for(j in 1:n){
    ij=(i-1)*n+j;
    u1[i,j]=u[ij];
    u2[i,j]=u[ij+n2];
    u3[i,j]=u[ij+n22];
  }
  }
```

Three arrays are also defined for the LHS derivatives in t of eqs. (5.1), u1t, u2t, u3t. The 1D-2D relation for the indices, ij=(i-1)*n+j, is used as in Listing 5.1.

- Arrays for the derivatives $\dfrac{\partial u_1}{\partial x}, \dfrac{\partial u_2}{\partial x}, \dfrac{\partial u_3}{\partial x}$ are declared (preallocated).

```
#
# u1x,u2x,u3x
  u1x=matrix(0,nrow=n,ncol=n);
  u2x=matrix(0,nrow=n,ncol=n);
  u3x=matrix(0,nrow=n,ncol=n);
```

- The first spatial derivatives, $\dfrac{\partial u_1}{\partial x}, \dfrac{\partial u_2}{\partial x}, \dfrac{\partial u_3}{\partial x}$, are computed with the spline utility splinefun.

```
  for(j in 1:n){
#
#   u1x
    table=splinefun(x,u1[,j]);
    u1x[,j]=table(x,deriv=1);
#
#   u2x
    table=splinefun(x,u2[,j]);
    u2x[,j]=table(x,deriv=1);
#
#   u3x
    table=splinefun(x,u3[,j]);
    u3x[,j]=table(x,deriv=1);
  }
```

deriv=1 specifies a first derivative is computed.
- Similarly, the first derivatives in y are computed with splinefun and placed in arrays u1y,u2y,u3y.

```
#
# u1y,u2y,u3y
```

```
u1y=matrix(0,nrow=n,ncol=n);
u2y=matrix(0,nrow=n,ncol=n);
u3y=matrix(0,nrow=n,ncol=n);
for(i in 1:n){
#
#   u1y
    table=splinefun(y,u1[i,]);
    u1y[i,]=table(y,deriv=1);
#
#   u2y
    table=splinefun(y,u2[i,]);
    u2y[i,]=table(y,deriv=1);
#
#   u3y
    table=splinefun(y,u3[i,]);
    u3y[i,]=table(y,deriv=1);
}
```

- The homogeneous BCs, eqs. (5.3), (5.4), are implemented

```
#
# BCs
  for(j in 1:n){
    u1x[1,j]=0;u1x[n,j]=0;
    u2x[1,j]=0;u2x[n,j]=0;
    u3x[1,j]=0;u3x[n,j]=0;
  }
  for(i in 1:n){
    u1y[i,1]=0;u1y[i,n]=0;
    u2y[i,1]=0;u2y[i,n]=0;
    u3y[i,1]=0;u3y[i,n]=0;
  }
```

Note the use of the subscripts 1,n for the boundaries in x and y.

- The second spatial derivatives, $\dfrac{\partial^2 u_1}{\partial x^2}, \dfrac{\partial^2 u_2}{\partial x^2}, \dfrac{\partial^2 u_3}{\partial x^2}$, are computed by differentiating the first derivatives (stage-wise differentiation).

```
#
# u1xx,u2xx,u3xx
  u1xx=matrix(0,nrow=n,ncol=n);
  u2xx=matrix(0,nrow=n,ncol=n);
  u3xx=matrix(0,nrow=n,ncol=n);
  for(j in 1:n){
#
#   u1xx
    table=splinefun(x,u1x[,j]);
    u1xx[,j]=table(x,deriv=1);
#
#   u2xx
    table=splinefun(x,u2x[,j]);
    u2xx[,j]=table(x,deriv=1);
#
#   u3xx
    table=splinefun(x,u3x[,j]);
    u3xx[,j]=table(x,deriv=1);
  }
```

- The second spatial derivatives, $\dfrac{\partial^2 u_1}{\partial y^2}, \dfrac{\partial^2 u_2}{\partial y^2}, \dfrac{\partial^2 u_3}{\partial y^2}$, are computed by differentiating the first derivatives.

```
#
# u1yy,u2yy,u3yy
  u1yy=matrix(0,nrow=n,ncol=n);
  u2yy=matrix(0,nrow=n,ncol=n);
```

```
    u3yy=matrix(0,nrow=n,ncol=n);
    for(i in 1:n){
#
#    u1yy
    table=splinefun(y,u1y[i,]);
    u1yy[i,]=table(y,deriv=1);
#
#    u2yy
    table=splinefun(y,u2y[i,]);
    u2yy[i,]=table(y,deriv=1);
#
#    u3yy
    table=splinefun(y,u3y[i,]);
    u3yy[i,]=table(y,deriv=1);
    }
```

- The MOL/ODEs of eqs. (5.1) are programmed over the full interval in x and y. The result is the set of ODE derivatives approximating $\dfrac{\partial u_1}{\partial t}, \dfrac{\partial u_2}{\partial t}, \dfrac{\partial u_3}{\partial t}$.

```
#
# u1t(x,y,t)
    for(i in 1:n){
    for(j in 1:n){
      u1t[i,j]=
        a11*u1[i,j]+a12*u2[i,j]+a13*u3[i,j]+
        D1*u1xx[i,j]+D1*u1yy[i,j];
      u2t[i,j]=
        a21*u1[i,j]+a22*u2[i,j]+a23*u3[i,j]+
        D2*u2xx[i,j]+D2*u2yy[i,j];
      u3t[i,j]=
        a31*u1[i,j]+a32*u2[i,j]+a33*u3[i,j]+
        D3*u3xx[i,j]+D3*u3yy[i,j];
#
```

```
# Next j
  }
#
# Next i
  }
```

- The three arrays with the derivatives in t are placed in one vector, ut (length $n3 = 3(21^2) = 1323$ set in the main program).

```
#
# Three arrays to one vector
  ut=rep(0,n3);
  for(i in 1:n){
  for(j in 1:n){
    ij=(i-1)*n+j;
    ut[ij]    =u1t[i,j];
    ut[ij+n2] =u2t[i,j];
    ut[ij+n22]=u3t[i,j];
  }
  }
```

- The number of calls to pde_1f is incremented.

```
#
# Increment calls to pde_1f
  ncall <<- ncall+1;
```

- The derivative vector ut is returned as a list as required by lsodes.

```
#
# Return derivative vector
  return(list(c(ut)));
  }
```

The final } concludes pde_1f.

(5.3.3) Model output

The numerical and graphical (plotted) output is considered next.

For ncase=4, the output from the main program and ODE/MOL routine of Listing 5.3 follows.

[1] 4

[1] 1324

```
t0 = 0.00
t1 = 6.67
t2 = 13.33
t3 = 20.00
```

```
ncall = 3334
```

Table 5.2: Numerical output for ncase=4, splines

We can note the following details about this output.

- As in the case of pde_1e, the solution array out from lsodes has the dimensions 4×1324, corresponding to $nout = 4$ and $3(21^2) + 1 = 1323 + 1$ with an offset +1 for the value of t as the first column element at each solution output point. Therefore, the ODE solution vector is in the column elements $2, \ldots, 1323 + 1$.
- The solution output points (values of t) correspond to t0=0, tf=20 as defined in the main program.

```
t0 = 0.00
t1 = 6.67
t2 = 13.33
t3 = 20.00
```

- The total calls to pde_1e is substantial, ncall = 3334.

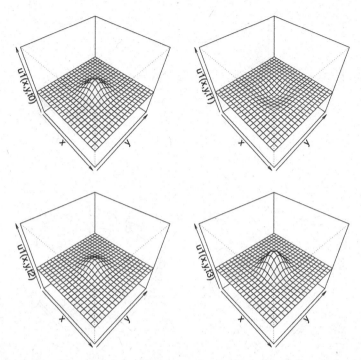

Figure 5.2a: Numerical solution of eq. (5.1a), $u_1(x, y, t)$, spline, ncase=4

The graphical output is in Figs. 5.2a ($u_1(x, y, t = 0, 6.67, 13.33, 20)$), 5.2b ($u_2(x, y, t = 0, 6.67, 13.33, 20)$), 5.2c ($u_3(x, y, t = 0, 6.67, 13.33, 20)$).

Figs. 5.1 and 5.2 are essentially identical, indicating that the FDs and spline MOL approximations produced the same numerical solution. This is not a proof that the solutions are correct, but the agreement implies good accuracy.

Similar results are produced with ncase=5, that is, the PDE solutions of eqs. (5.1) oscillate rather than a monotonic decay (dispersion) from diffusion only.

In summary, eqs. (5.1) again demonstrate oscillation for ncase=4 with the parameters orginally proposed by Turing.

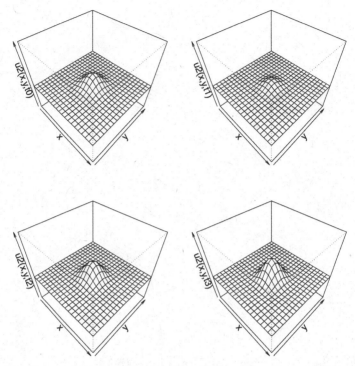

Figure 5.2b: Numerical solution of eq. (5.1b), $u_2(x, y, t)$, spline, ncase=4

(5.4) Summary and Conclusions

The extension from 1D to 2D with finite difference MOL, as programmed in pde_1e of Listing 5.2, and splines in pde_1f in Listing 5.3 is straightforward. The gridding in x, y with 21 points in each dimension gives acceptable spatial resolution for ncase=4,5 (if random ICs rather than the Gaussian function are used, the number of points would probably have to be increased, but this would have to be done with caution since the number of approximating MOL/ODEs increases as the product of the number of points in x and y).

Parameter variations beyond ncase=1,2,3,4,5 can be studied as extensions of the RD PDE model (with additions to the

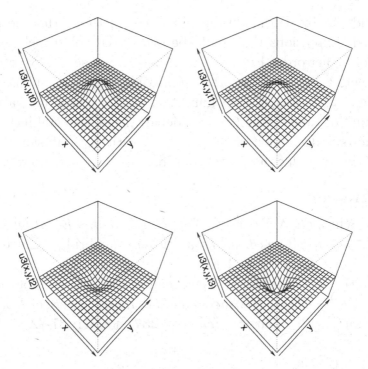

Figure 5.2c: Numerical solution of eq. (5.1c), $u_3(x, y, t)$, splines, ncase=4

main program of Listing 5.1). Other ICs and BCs could be easily investigated with modification of Listings 5.1, 5.2, 5.3 e.g., Dirichlet, Robin BCs. Also, the form of the PDEs (eq. (5.1)) can be varied, e.g., to include nonlinear chemotaxis.

Thus, the computer implementation with the R routines offers the possibility of experimentation with alternative RD models. As a word of caution, the performance of modified R routines cannot be guaranteed in advance, and some experimentation with the coding and parameters, such as the intervals in x, y, t, may be required.

Finally, the overall intent of the book is to provide some explanation of and insight into the morphogenesis RD PDEs considered by Turing, using numerical methods that he suggested. Extensions that could be considered could include the use of higher order FD approximations (discussed in Chapter 5, Appendix A1), or some other approximations (e.g., splines discussed in Chapter 5, Appendix A2), for the spatial derivatives. Oscillation and stability that depart from the stability and monotonic dispersion of PDE systems with just diffusion would be features of the numerical solutions of particular interest.

References

[1]   Schiesser, W.E. (2013), *Partial Differential Equation Analysis in Biomedical Engineering*, Cambridge University Press, Cambridge, UK

[2]   Turing, A.M. (1952), The chemical basis of morphogenesis, *Philosophical Transactions of the Royal Society of London, Series B, Biological Sciences*, **237**, no. 641, 37-72

Appendix A1: Spatial Approximations
with Finite Differences

In the preceding MOL solutions of the reaction-diffusion (RD) PDEs, the spatial derivatives were approximated by finite differences (FDs) of second order[1] which suggests that the numerical solution may have limited accuracy. The question of accuracy can be studied in two ways without requiring an analytical solution (which is usually unavailable in most PDE applications).

- *h refinement* in which the number of spatial grid points is varied and the accuracy inferred by comparing solutions with differing numbers of grid points.
- *p refinement* in which the order of the approximation of the spatial derivatives is varied and the accuracy is inferred by comparing solutions with differing orders.

---

[1]The order of a FD approximation pertains to the power (order) of the error resulting from the use of a finite interval in the spatial (independent) variable. For example, if the spatial variable is x, the truncation error is given by

$$err = c\Delta x^p$$

where p is the order of the approximation. For the FDs used in the preceding PDE examples, $p = 2$.

Here we consider an application of p refinement by using FDs of varying orders. This is done by calling FD library routines (rather than program the FDs which increase in complexity as the order increases).

(A1.1) R Routines

The MOL solution of eqs. (1.5) is computed with FDs of varying order by the following R routines, starting with a main program.

(A1.1.1) Main program

```
#
# 1D 3 x 3 Turing, FDs
#
# Delete previous workspaces
  rm(list=ls(all=TRUE))
#
# Access ODE integrator
  library("deSolve");
#
# Access functions for numerical solution
  setwd("f:/turing/1D");
  source("pde_A1.R");
  source("dss042.R");
  source("dss044.R");
  source("dss046.R");
#
# Grid in x
  xl=0;xu=1;n=51;dx=(xu-xl)/(n-1);
  x=seq(from=xl,to=xu,by=dx);
  n2=2*n;n3=3*n;
#
# Parameters
  ncase=4;
```

```
if(ncase==1){
  a11  =0; a12  =0; a13  =0;
  a21  =0; a22  =0; a23  =0;
  a31  =0; a32  =0; a33  =0;
  D1=dx^2; D2=dx^2; D3=dx^2;
  t0   =0; tf  =20; nout =6;
  c1   =0; c2  =50;
  zl   =0; zu   =1;}
if(ncase==2){
  a11= -1; a12  =0; a13  =0;
  a21 = 0; a22 =-1; a23 = 0;
  a31 = 0; a32  =0; a33 =-1;
  D1=dx^2; D2=dx^2; D3=dx^2;
  t0   =0; tf  =20; nout =6;
  c1   =0; c2  =50;
  zl   =0; zu   =1;}
if(ncase==3){
  a11  =1; a12  =0; a13  =0;
  a21  =0; a22  =1; a23  =0;
  a31  =0; a32  =0; a33  =1;
  D1=dx^2; D2=dx^2; D3=dx^2;
  t0   =0; tf   =2; nout =6;
  c1   =0; c2  =50;
  zl   =0; zu   =1;}
if(ncase==4){
a11  =-10/3; a12      =3; a13   =-1;
a21      =-2; a22   =7/3; a23    =0;
a31       =3; a32     =-4; a33    =0;
D1=2/3*dx^2; D2=1/3*dx^2; D3=0*dx^2;
t0        =0; tf      =20; nout    =6;
c1        =0; c2      =50;
zl       =-2; zu       =3;}
if(ncase==5){
a11   =-1; a12    =-1; a13   = 0;
```

```
  a21    =1; a22    =0; a23   =-1;
  a31    =0; a32    =1; a33    =0;
  D1=1*dx^2; D2=0*dx^2; D3=0*dx^2;
  t0     =0; tf     =20; nout   =6;
  c1     =0; c2     =50;
  zl    =-1; zu     =2;}
#
# Independent variable for ODE integration
  tout=seq(from=t0,to=tf,by=(tf-t0)/(nout-1));
#
# ICs
  u0=rep(0,n3);
  for(i in 1:n){
    u0[i]   =c1+exp(-c2*(x[i]-0.5)^2);
    u0[i+n] =c1+exp(-c2*(x[i]-0.5)^2);
    u0[i+n2]=c1+exp(-c2*(x[i]-0.5)^2);
#   u0[i+n] =0;
#   u0[i+n2]=0;
  }
  ncall=0;
#
# ODE integration
  out=lsodes(y=u0,times=tout,func=pde_A1,
    sparsetype="sparseint",rtol=1e-6,
    atol=1e-6,maxord=5);
  nrow(out)
  ncol(out)
#
# Arrays for numerical solution
  u1=matrix(0,nrow=n,ncol=nout);
  u2=matrix(0,nrow=n,ncol=nout);
  u3=matrix(0,nrow=n,ncol=nout);
  t=rep(0,nout);
  for(it in 1:nout){
```

```
    for(i  in 1:n){
      u1[i,it]=out[it,i+1];
      u2[i,it]=out[it,i+1+n];
      u3[i,it]=out[it,i+1+n2];
        t[it]=out[it,1];
    }
    }
#
# Display selected output
    for(it in 1:nout){
      cat(sprintf("\n      t          x    u1(x,t)
                  u2(x,t)   u3(x,t)\n"));
      iv=seq(from=1,to=n,by=5);
      for(i in iv){
        cat(sprintf(
          "%6.2f%9.3f%10.6f%10.6f%10.6f\n",
          t[it],x[i],u1[i,it],u2[i,it],u3[i,it]));
        }
        cat(sprintf("\n"));
      }
    cat(sprintf(" ncall = %4d\n",ncall));
#
# Plot 2D numerical solution
    matplot(x,u1,type="l",lwd=2,col="black",lty=1,
      xlab="x",ylab="u1(x,t)",main="");
    matplot(x,u2,type="l",lwd=2,col="black",lty=1,
      xlab="x",ylab="u2(x,t)",main="");
    matplot(x,u3,type="l",lwd=2,col="black",lty=1,
      xlab="x",ylab="u3(x,t)",main="");
#
# Plot 3D numerical solution
    persp(x,t,u1,theta=45,phi=45,xlim=c(xl,xu),
        ylim=c(t0,tf),xlab="x",ylab="t",
        zlab="u1(x,t)");
```

```
persp(x,t,u2,theta=45,phi=45,xlim=c(xl,xu),
    ylim=c(t0,tf),xlab="x",ylab="t",
    zlab="u2(x,t)");
persp(x,t,u3,theta=45,phi=45,xlim=c(xl,xu),
    ylim=c(t0,tf),xlab="x",ylab="t",
    zlab="u3(x,t)");
```

Listing A1.1: Main program for eqs. (1.5)

Listing A1.1 is similar to Listing 1.1, so only the differences are explained.

- In addition to the ODE/MOL routine pde_A1, three library FD routines are accessed, dss042, dss044, dss046 [1].

```
#
# 1D 3 x 3 Turing, FDs
#
# Delete previous workspaces
    rm(list=ls(all=TRUE))
#
# Access ODE integrator
    library("deSolve");
#
# Access functions for numerical solution
    setwd("f:/turing/1D");
    source("pde_A1.R");
    source("dss042.R");
    source("dss044.R");
    source("dss046.R");
```

- The three factors in Listing 1.1

```
D1dx2,D2dx2,D3dx2
```

  are not required as explained in the subsequent discussion of pde_A1.

- pde_A1 is used in the ODE integration.

```
#
# ODE integration
  out=lsodes(y=u0,times=tout,func=pde_A1,
      sparsetype="sparseint",rtol=1e-6,
      atol=1e-6,maxord=5);
  nrow(out)
  ncol(out)
```

Otherwise, the main programs in Listings 1.1 and A1.1 are the same.

(A1.1.2) ODE/MOL routine

The ODE/MOL routine called in the main program of Listing A1.1 follows.

```
  pde_A1=function(t,u,parm){
#
# Function pde_A1 computes the t derivative
# vector for u1(x,t),u2(x,t),u3(x,t)
#
# One vector to three vectors
  u1=rep(0,n); u2=rep(0,n); u3=rep(0,n);
  u1t=rep(0,n);u2t=rep(0,n);u3t=rep(0,n);
  for(i in 1:n){
    u1[i]=u[i];
    u2[i]=u[i+n];
    u3[i]=u[i+n2]
  }
#
# BCs
  nl=2;nu=2;
  u1x=rep(0,n);u2x=rep(0,n);u3x=rep(0,n);
  u1x[1]=0;u1x[n]=0;
```

```
  u2x[1]=0;u2x[n]=0;
  u3x[1]=0;u3x[n]=0;
#
# u1xx
  u1xx=dss042(xl,xu,n,u1,u1x,nl,nu);
# u1xx=dss044(xl,xu,n,u1,u1x,nl,nu);
# u1xx=dss046(xl,xu,n,u1,u1x,nl,nu);
#
# u2xx
  u2xx=dss042(xl,xu,n,u2,u2x,nl,nu);
# u2xx=dss044(xl,xu,n,u2,u2x,nl,nu);
# u2xx=dss046(xl,xu,n,u2,u2x,nl,nu);
#
# u3xx
  u3xx=dss042(xl,xu,n,u3,u3x,nl,nu);
# u3xx=dss044(xl,xu,n,u3,u3x,nl,nu);
# u3xx=dss046(xl,xu,n,u3,u3x,nl,nu);
#
# u1t(x,t)
  for(i in 1:n){
    u1t[i]=a11*u1[i]+a12*u2[i]+a13*u3[i]+
          D1*u1xx[i];
  }
#
# u2t(x,t)
  for(i in 1:n){
    u2t[i]=a21*u1[i]+a22*u2[i]+a23*u3[i]+
          D2*u2xx[i];
  }
#
# u3t(x,t)
  for(i in 1:n){
    u3t[i]=a31*u1[i]+a32*u2[i]+a33*u3[i]+
          D3*u3xx[i];
```

```
}
#
# Three vectors to one vector
  ut=rep(0,n3);
  for(i in 1:n){
    ut[i]    =u1t[i];
    ut[i+n]  =u2t[i];
    ut[i+n2]=u3t[i];
  }
#
# Increment calls to pde_A1
  ncall <<- ncall+1;
#
# Return derivative vector
  return(list(c(ut)));
  }
```

Listing A1.2: ODE/MOL routine pde_A1 for eqs. (1.5)

We can note the following details about pde_A1.

- The function is defined.

```
  pde_A1=function(t,u,parm){
  #
  # Function pde_A1 computes the t derivative
  # vector for u1(x,t),u2(x,t),u3(x,t)
```

t is the current value of t in eqs. (1.5). u is the 153-vector of ODE/MOL dependent variables. parm is an argument to pass parameters to pde_A1 (unused, but required in the argument list). The arguments must be listed in the order stated to properly interface with lsodes called in the main program of Listing A1.1. The composite derivative vector of the LHSs of eqs. (1.5) is calculated next and returned to lsodes.

- The dependent variable vectors are placed in three vectors to facilitate the programming of eqs. (1.5).

```
#
# One vector to three vectors
  u1=rep(0,n); u2=rep(0,n); u3=rep(0,n);
  u1t=rep(0,n);u2t=rep(0,n);u3t=rep(0,n);
  for(i in 1:n){
    u1[i]=u[i];
    u2[i]=u[i+n];
    u3[i]=u[i+n2]
  }
```

Vectors are also defined for the LHS derivatives in t of eqs. (1.5).

- Homogeneous Neumann BCs are applied at $x = x_l, x_u$.

$$\frac{\partial u_1(x = x_l, t)}{\partial x} = \frac{\partial u_2(x = x_l, t)}{\partial x} = \frac{\partial u_3(x = x_l, t)}{\partial x} = 0$$
(A1.1a,b,c)

$$\frac{\partial u_1(x = x_u, t)}{\partial x} = \frac{\partial u_2(x = x_u, t)}{\partial x} = \frac{\partial u_3(x = x_u, t)}{\partial x} = 0$$
(A1.1d,e,f)

```
#
# BCs
  nl=2;nu=2;
  u1x=rep(0,n);u2x=rep(0,n);u3x=rep(0,n);
  u1x[1]=0;u1x[n]=0;
  u2x[1]=0;u2x[n]=0;
  u3x[1]=0;u3x[n]=0;
```

nl=nu=2 specify Neumann BCs (nl=nu=1 would specify Dirichlet BCs).

- The second order spatial derivatives $\frac{\partial^2 u_1}{\partial x^2}, \frac{\partial^2 u_2}{\partial x^2}, \frac{\partial^2 u_3}{\partial x^2}$, are computed by a library differentiation routine selected by uncommenting particular lines. To start, dss042 is used which is based on the same three point, second order FDs as explictly programmed in Listing 1.2.

```
#
# u1xx
  u1xx=dss042(xl,xu,n,u1,u1x,nl,nu);
# u1xx=dss044(xl,xu,n,u1,u1x,nl,nu);
# u1xx=dss046(xl,xu,n,u1,u1x,nl,nu);
#
# u2xx
  u2xx=dss042(xl,xu,n,u2,u2x,nl,nu);
# u2xx=dss044(xl,xu,n,u2,u2x,nl,nu);
# u2xx=dss046(xl,xu,n,u2,u2x,nl,nu);
#
# u3xx
  u3xx=dss042(xl,xu,n,u3,u3x,nl,nu);
# u3xx=dss044(xl,xu,n,u3,u3x,nl,nu);
# u3xx=dss046(xl,xu,n,u3,u3x,nl,nu);
```

dss044 and dss046 are based on five point, fourth order and seven point, sixth order FDs, respectively.

- Eqs. (1.5) are programmed over the interval $x_l \leq x \leq x_u$.

```
#
# u1t(x,t)
  for(i in 1:n){
    u1t[i]=a11*u1[i]+a12*u2[i]+a13*u3[i]+
           D1*u1xx[i];
  }
#
# u2t(x,t)
  for(i in 1:n){
    u2t[i]=a21*u1[i]+a22*u2[i]+a23*u3[i]+
           D2*u2xx[i];
```

```
    }
#
# u3t(x,t)
  for(i in 1:n){
    u3t[i]=a31*u1[i]+a32*u2[i]+a33*u3[i]+
           D3*u3xx[i];
    }
```

The result is the set of MOL/ODEs approximating eqs. (1.5). u1xx, u2xx, u3xx include a division by Δx^2 so this division is not included in the main program of Listing A1.1 (factors D1dx2,D2dx2,D3dx2 are not defined in Listing A1.1 as they are in Listing 1.1).

- The three vectors u1t,u2t,u3t are placed in a single vector u of length $3(51) = 153$ for return to lsodes.

```
#
# Three vectors to one vector
  ut=rep(0,n3);
  for(i in 1:n){
    ut[i]    =u1t[i];
    ut[i+n]  =u2t[i];
    ut[i+n2]=u3t[i];
    }
```

- The counter for the calls to **pde_A1** is incremented and its value is returned to the calling program (of Listing A1.1) with the <<- operator.

```
#
# Increment calls to pde_A1
  ncall <<- ncall+1;
```

- The derivative vector (LHSs of eqs. (1.5)) is returned to lsodes which requires a list. c is the R vector utility. The combination of **return, list, c** gives lsodes (the

ODE integrator called in the main program of Listing A1.1) the required derivative vector for the next step along the solution.

```
#
# Return .derivative vector
  return(list(c(ut)));
  }
```

The final } concludes pde_A1.

The numerical and graphical (plotted) output is considered next.

(A1.2) Model Output

Abbreviated numerical output for **ncase=4** follows.

dss042

[1] 6

[1] 154

t	x	u1(x,t)	u2(x,t)	u3(x,t)
0.00	0.000	0.000004	0.000004	0.000004
0.00	0.100	0.000335	0.000335	0.000335
0.00	0.200	0.011109	0.011109	0.011109
0.00	0.300	0.135335	0.135335	0.135335
0.00	0.400	0.606531	0.606531	0.606531
0.00	0.500	1.000000	1.000000	1.000000
0.00	0.600	0.606531	0.606531	0.606531
0.00	0.700	0.135335	0.135335	0.135335
0.00	0.800	0.011109	0.011109	0.011109
0.00	0.900	0.000335	0.000335	0.000335
0.00	1.000	0.000004	0.000004	0.000004

.

.

.

Output for t = 4,...,16 removed

.

.

.

t	x	u1(x,t)	u2(x,t)	u3(x,t)
20.00	0.000	-0.000034	-0.000022	0.000038
20.00	0.100	-0.000218	-0.000321	0.000454
20.00	0.200	0.007326	0.001532	-0.005541
20.00	0.300	0.144030	0.091331	-0.162128
20.00	0.400	0.723723	0.560826	-0.877270
20.00	0.500	1.224314	0.998136	-1.508758
20.00	0.600	0.723723	0.560826	-0.877270
20.00	0.700	0.144030	0.091331	-0.162128
20.00	0.800	0.007326	0.001532	-0.005541
20.00	0.900	-0.000218	-0.000321	0.000454
20.00	1.000	-0.000034	-0.000022	0.000038

ncall =　363

Table A1.1: Abbreviated numerical output for **ncase=4**, dss042

The abbreviated numerical output for the explicit FD (Turing) MOL/ODEs from Chapter 1 is repeated for comparison.

Explicit FD (Turing)

[1] 6

[1] 154

t	x	u1(x,t)	u2(x,t)	u3(x,t)
0.00	0.000	0.000004	0.000004	0.000004

```
0.00     0.100   0.000335   0.000335   0.000335
0.00     0.200   0.011109   0.011109   0.011109
0.00     0.300   0.135335   0.135335   0.135335
0.00     0.400   0.606531   0.606531   0.606531
0.00     0.500   1.000000   1.000000   1.000000
0.00     0.600   0.606531   0.606531   0.606531
0.00     0.700   0.135335   0.135335   0.135335
0.00     0.800   0.011109   0.011109   0.011109
0.00     0.900   0.000335   0.000335   0.000335
0.00     1.000   0.000004   0.000004   0.000004
                    .          .          .
                    .          .          .
                    .          .          .

         Output for t = 4,...,16 removed
                    .          .
                    .          .
                    .          .

    t        x     u1(x,t)     u2(x,t)     u3(x,t)
20.00     0.000   -0.000035   -0.000024    0.000041
20.00     0.100   -0.000218   -0.000321    0.000454
20.00     0.200    0.007326    0.001532   -0.005541
20.00     0.300    0.144030    0.091331   -0.162128
20.00     0.400    0.723723    0.560826   -0.877270
20.00     0.500    1.224314    0.998136   -1.508758
20.00     0.600    0.723723    0.560826   -0.877270
20.00     0.700    0.144030    0.091331   -0.162128
20.00     0.800    0.007326    0.001532   -0.005541
20.00     0.900   -0.000218   -0.000321    0.000454
20.00     1.000   -0.000035   -0.000024    0.000041

ncall =  363
```

Table A1.2: Abbreviated numerical output for **ncase=4**,
explicit FDs

The output in Tables A1.1 and A1.2 is essentially the same indicating that the three point, second order FDs of dss042 are the same as the explicit FDs stated by Turing.

The 3D plots from persp are in Figs. A1.1a, A1.1b, A1.1c.

Figs. A1.1a,b,c are the same as Figs. 1.2c,d,e (Chapter 1) as expected (because of the close repetition in Tables A1.1 and A1.2).

If the routines of Listings A1.1, A1.2 are executed with dss042, dss044, dss046, the following abbreviated output results (for $t = 20$). Also included in Table A1.3 is the abbreviated output for the explicit FDs from Chapter 1, and the splines discussed next in Appendix A2, so that the output from these various MOL formulations of eqs. (1.5) is in one place to facilitate comparison.

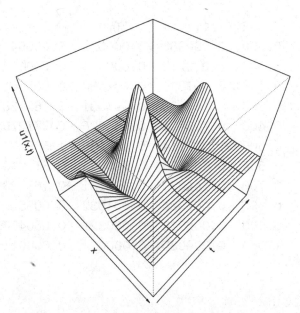

Figure A1.1a: Numerical solution of eq. (1.5a) for $u_1(x,t)$, dss042, ncase=4

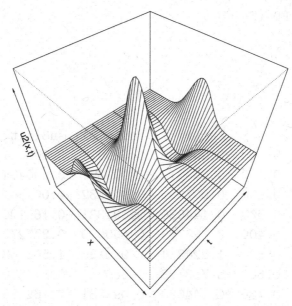

Figure A1.1b: Numerical solution of eq. (1.5b) for $u_2(x,t)$, dss042, ncase=4

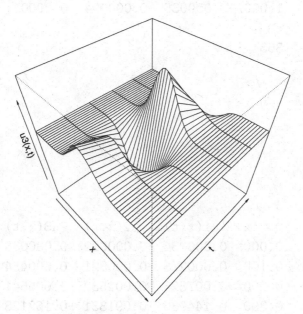

Figure A1.1c: Numerical solution of eq. (1.5c) for $u_3(x,t)$, dss042, ncase=4

Explicit FD (Turing)

[1] 6

[1] 154

t	x	u1(x,t)	u2(x,t)	u3(x,t)
20.00	0.000	-0.000035	-0.000024	0.000041
20.00	0.100	-0.000218	-0.000321	0.000454
20.00	0.200	0.007326	0.001532	-0.005541
20.00	0.300	0.144030	0.091331	-0.162128
20.00	0.400	0.723723	0.560826	-0.877270
20.00	0.500	1.224314	0.998136	-1.508758
20.00	0.600	0.723723	0.560826	-0.877270
20.00	0.700	0.144030	0.091331	-0.162128
20.00	0.800	0.007326	0.001532	-0.005541
20.00	0.900	-0.000218	-0.000321	0.000454
20.00	1.000	-0.000035	-0.000024	0.000041

ncall =   363

dss042

[1] 6

[1] 154

t	x	u1(x,t)	u2(x,t)	u3(x,t)
20.00	0.000	-0.000034	-0.000022	0.000038
20.00	0.100	-0.000218	-0.000321	0.000454
20.00	0.200	0.007326	0.001532	-0.005541
20.00	0.300	0.144030	0.091331	-0.162128
20.00	0.400	0.723723	0.560826	-0.877270

20.00	0.500	1.224314	0.998136	-1.508758
20.00	0.600	0.723723	0.560826	-0.877270
20.00	0.700	0.144030	0.091331	-0.162128
20.00	0.800	0.007326	0.001532	-0.005541
20.00	0.900	-0.000218	-0.000321	0.000454
20.00	1.000	-0.000034	-0.000022	0.000038

ncall = 363

dss044

[1] 6

[1] 154

t	x	u1(x,t)	u2(x,t)	u3(x,t)
20.00	0.000	-0.000031	-0.000022	0.000036
20.00	0.100	-0.000201	-0.000299	0.000422
20.00	0.200	0.007336	0.001603	-0.005578
20.00	0.300	0.143931	0.091125	-0.161922
20.00	0.400	0.723694	0.560626	-0.877199
20.00	0.500	1.224511	0.998759	-1.509171
20.00	0.600	0.723694	0.560626	-0.877199
20.00	0.700	0.143931	0.091125	-0.161922
20.00	0.800	0.007336	0.001603	-0.005578
20.00	0.900	-0.000201	-0.000299	0.000422
20.00	1.000	-0.000031	-0.000022	0.000036

ncall = 383

dss046

[1] 6

[1] 154

```
    t         x     u1(x,t)    u2(x,t)    u3(x,t)
20.00     0.000  -0.000031  -0.000021   0.000035
20.00     0.100  -0.000201  -0.000300   0.000422
20.00     0.200   0.007337   0.001605  -0.005580
20.00     0.300   0.143930   0.091125  -0.161920
20.00     0.400   0.723692   0.560618  -0.877196
20.00     0.500   1.224514   0.998773  -1.509178
20.00     0.600   0.723692   0.560618  -0.877196
20.00     0.700   0.143930   0.091125  -0.161920
20.00     0.800   0.007337   0.001605  -0.005580
20.00     0.900  -0.000201  -0.000300   0.000422
20.00     1.000  -0.000031  -0.000021   0.000035
```

ncall = 399

splinefun (Appendix A2)

[1] 6

[1] 154

```
    t         x     u1(x,t)    u2(x,t)    u3(x,t)
20.00     0.000  -0.000032  -0.000022   0.000037
20.00     0.100  -0.000201  -0.000299   0.000422
20.00     0.200   0.007336   0.001603  -0.005578
20.00     0.300   0.143931   0.091125  -0.161922
20.00     0.400   0.723694   0.560627  -0.877199
20.00     0.500   1.224510   0.998758  -1.509170
20.00     0.600   0.723694   0.560627  -0.877199
```

```
20.00     0.700   0.143931   0.091125  -0.161922
20.00     0.800   0.007336   0.001603  -0.005578
20.00     0.900  -0.000201  -0.000299   0.000422
20.00     1.000  -0.000032  -0.000022   0.000037


ncall =   615
```

Table A1.3: Abbreviated numerical output for **ncase=4**, explicit FDs, **dss042, dss044, dss046**, splines in **splinefun**

Table A1.3 indicates the various MOL formulations agree to 4+ figures.

(A1.3) Summary and Conclusions

The MOL numerical integration of eqs. (1.5) with homogeneous Neumann BCs is straightforward. Variations of the PDEs could now be easily investigated, e.g., Dirichlet, Robin BCs, nonlinear terms in the PDEs, addition of PDEs for more than three components. Also, computing and displaying the RHS and LHS terms of eqs. (1.5) would give additional insight into features of the solutions of these equations.

The preceding variation in the MOL formulation demonstrates a form of p refinement with: (1) variation in the order of the approximations of the spatial derivatives through **dss042, dss044, dss046** and (2) variation in the approximations through explicit FDs (from Turing), the FD **dss** routines and spline collocation with **splinefun**. The close agreement between the solutions does not constitute a proof of correctness and/or numerical accuracy of the solutions, and the agreement only implies or infers good accuracy. But this form of error analysis (plus h refinement in which the number of spatial points is varied) does not require an analytical solution, and should be part of any newly reported numerical PDE solution.

References

[1] Schiesser, W.E. (2013), *Partial Differential Equation Analysis in Biomedical Engineering*, Cambridge University Press, Cambridge, UK

[2] Turing, A.M. (1952), The chemical basis of morphogenesis, *Philosophical Transactions of the Royal Society of London, Series B, Biological Sciences*, **237**, no. 641, 37-72

Appendix A2: Spatial Approximations with Splines

In the preceding MOL solutions of the reaction-diffusion (RD) PDEs, the spatial derivatives were approximated by finite differences (FDs) of varying order (from dss042, dss044, dss046). In this appendix, we consider the MOL solution of eqs. (1.5) with the spatial derivatives in x approximated with splines.

(A2.1) R Routines

The MOL solution of eqs. (1.5) is computed with splines by the following R routines, starting with a main program.

(A2.1.1) Main program

```
#
# 1D 3 x 3 Turing, splines
#
# Delete previous workspaces
  rm(list=ls(all=TRUE))
#
# Access ODE integrator
  library("deSolve");
#
# Access functions for numerical solution
  setwd("f:/turing/1D");
```

```
   source("pde_A2.R");
#
# Grid in x
   xl=0;xu=1;n=51;dx=(xu-xl)/(n-1);
   x=seq(from=xl,to=xu,by=dx);
   n2=2*n;n3=3*n;
#
# Parameters
   ncase=4;
   if(ncase==1){
     a11  =0; a12  =0; a13  =0;
     a21  =0; a22  =0; a23  =0;
     a31  =0; a32  =0; a33  =0;
     D1=dx^2; D2=dx^2; D3=dx^2;
     t0   =0; tf  =20; nout =6;
     c1   =0; c2  =50;
     zl   =0; zu   =1;}
   if(ncase==2){
     a11= -1; a12  =0; a13  =0;
     a21 = 0; a22 =-1; a23 = 0;
     a31 = 0; a32  =0; a33 =-1;
     D1=dx^2; D2=dx^2; D3=dx^2;
     t0   =0; tf  =20; nout =6;
     c1   =0; c2  =50;
     zl   =0; zu   =1;}
   if(ncase==3){
     a11  =1; a12  =0; a13  =0;
     a21  =0; a22  =1; a23  =0;
     a31  =0; a32  =0; a33  =1;
     D1=dx^2; D2=dx^2; D3=dx^2;
     t0   =0; tf   =2; nout =6;
     c1   =0; c2  =50;
     zl   =0; zu   =1;}
   if(ncase==4){
```

```
a11  =-10/3; a12        =3; a13   =-1;
a21        =-2; a22     =7/3; a23   =0;
a31        =3; a32        =-4; a33   =0;
D1=2/3*dx^2; D2=1/3*dx^2; D3=0*dx^2;
t0        =0; tf        =20; nout   =6;
c1        =0; c2        =50;
zl        =-2; zu        =3;}
if(ncase==5){
a11    =-1; a12    =-1; a13    = 0;
a21    =1; a22     =0; a23    =-1;
a31    =0; a32     =1; a33    =0;
D1=1*dx^2; D2=0*dx^2; D3=0*dx^2;
t0     =0; tf     =20; nout    =6;
c1     =0; c2     =50;
zl     =-1; zu     =2;}
#
# Independent variable for ODE integration
tout=seq(from=t0,to=tf,by=(tf-t0)/(nout-1));
#
# ICs
u0=rep(0,n3);
for(i in 1:n){
  u0[i]    =c1+exp(-c2*(x[i]-0.5)^2);
  u0[i+n]  =c1+exp(-c2*(x[i]-0.5)^2);
  u0[i+n2]=c1+exp(-c2*(x[i]-0.5)^2);
#   u0[i+n]  =0;
#   u0[i+n2]=0;
}
ncall=0;
#
# ODE integration
out=lsodes(y=u0,times=tout,func=pde_A2,
    sparsetype="sparseint",rtol=1e-6,
    atol=1e-6,maxord=5);
```

```
  nrow(out)
  ncol(out)
#
# Arrays for numerical solution
  u1=matrix(0,nrow=n,ncol=nout);
  u2=matrix(0,nrow=n,ncol=nout);
  u3=matrix(0,nrow=n,ncol=nout);
  t=rep(0,nout);
  for(it in 1:nout){
  for(i  in 1:n){
    u1[i,it]=out[it,i+1];
    u2[i,it]=out[it,i+1+n];
    u3[i,it]=out[it,i+1+n2];
      t[it]=out[it,1];
  }
  }
#
# Display selected output
  for(it in 1:nout){
    cat(sprintf("\n        t           x    u1(x,t)
              u2(x,t)    u3(x,t)\n"));
    iv=seq(from=1,to=n,by=5);
    for(i in iv){
     cat(sprintf(
       "%6.2f%9.3f%10.6f%10.6f%10.6f\n",
       t[it],x[i],u1[i,it],u2[i,it],u3[i,it]));
     }
     cat(sprintf("\n"));
   }
  cat(sprintf(" ncall = %4d\n",ncall));
#
# Plot 2D numerical solution
  matplot(x,u1,type="l",lwd=2,col="black",lty=1,
    xlab="x",ylab="u1(x,t)",main="");
```

```
matplot(x,u2,type="l",lwd=2,col="black",lty=1,
   xlab="x",ylab="u2(x,t)",main="");
matplot(x,u3,type="l",lwd=2,col="black",lty=1,
   xlab="x",ylab="u3(x,t)",main="");
#
# Plot 3D numerical solution
 persp(x,t,u1,theta=45,phi=45,xlim=c(xl,xu),
        ylim=c(t0,tf),xlab="x",ylab="t",
        zlab="u1(x,t)");
 persp(x,t,u2,theta=45,phi=45,xlim=c(xl,xu),
        ylim=c(t0,tf),xlab="x",ylab="t",
        zlab="u2(x,t)");
 persp(x,t,u3,theta=45,phi=45,xlim=c(xl,xu),
        ylim=c(t0,tf),xlab="x",ylab="t",
        zlab="u3(x,t)");
```

Listing A2.1: Main program for eqs. (1.5)

Listing A2.1 is similar to Listing 1.1, so only the differences are explained.

- The ODE/MOL routine is pde_A2 (discussed subsequently).

```
#
# 1D 3 x 3 Turing, splines
#
# Delete previous workspaces
  rm(list=ls(all=TRUE))
#
# Access ODE integrator
  library("deSolve");
#
# Access functions for numerical solution
  setwd("f:/turing/1D");
  source("pde_A2.R");
```

- The three factors in Listing 1.1

  ```
  D1dx2,D2dx2,D3dx2
  ```

  are not required as explained in the subsequent discussion of pde_A2.
- pde_A2 is used in the ODE integration.

```
#
# ODE integration
  out=lsodes(y=u0,times=tout,func=pde_A2,
    sparsetype="sparseint",rtol=1e-6,
    atol=1e-6,maxord=5);
  nrow(out)
  ncol(out)
```

Otherwise, the main programs in Listings 1.1 and A2.1 are the same.

(A2.1.2) ODE/MOL routine

The ODE/MOL routine called in the main program of Listing A2.1 follows.

```
  pde_A2=function(t,u,parm){
#
# Function pde_A2 computes the t derivative
# vector for u1(x,t),u2(x,t),u3(x,t)
#
# One vector to three vectors
  u1=rep(0,n); u2=rep(0,n); u3=rep(0,n);
  u1t=rep(0,n);u2t=rep(0,n);u3t=rep(0,n);
  for(i  in 1:n){
    u1[i]=u[i];
    u2[i]=u[i+n];
    u3[i]=u[i+n2]
  }
```

```
#
# u1x
  tablex=splinefun(x,u1);
  u1x=tablex(x,deriv=1);
#
# u2x
  tablex=splinefun(x,u2);
  u2x=tablex(x,deriv=1);
#
# u3x
  tablex=splinefun(x,u3);
  u3x=tablex(x,deriv=1);
#
# BCs
  u1x[1]=0;u1x[n]=0;
  u2x[1]=0;u2x[n]=0;
  u3x[1]=0;u3x[n]=0;
#
# u1xx
  tablexx=splinefun(x,u1x);
  u1xx=tablexx(x,deriv=1);
#
# u2xx
  tablexx=splinefun(x,u2x);
  u2xx=tablexx(x,deriv=1);
#
# u3xx
  tablexx=splinefun(x,u3x);
  u3xx=tablexx(x,deriv=1);
#
# u1t(x,t)
  for(i in 1:n){
    u1t[i]=a11*u1[i]+a12*u2[i]+a13*u3[i]+
           D1*u1xx[i];
```

```
  }
#
# u2t(x,t)
  for(i in 1:n){
    u2t[i]=a21*u1[i]+a22*u2[i]+a23*u3[i]+
          D2*u2xx[i];
  }
#
# u3t(x,t)
  for(i in 1:n){
    u3t[i]=a31*u1[i]+a32*u2[i]+a33*u3[i]+
          D3*u3xx[i];
  }
#
# Three vectors to one vector
  ut=rep(0,n3);
  for(i in 1:n){
    ut[i]    =u1t[i];
    ut[i+n]  =u2t[i];
    ut[i+n2]=u3t[i];
  }
#
# Increment calls to pde_A2
  ncall <<- ncall+1;
#
# Return derivative vector
  return(list(c(ut)));
  }
```

We can note the following details about **pde_A2**.

- The function is defined.

```
    pde_A2=function(t,u,parm){
  #
```

```
# Function pde_A2 computes the t derivative
# vector for u1(x,t),u2(x,t),u3(x,t)
```

t is the current value of *t* in eqs. (1.5). u is the 153-vector of ODE/MOL dependent variables. parm is an argument to pass parameters to pde_A2 (unused, but required in the argument list). The arguments must be listed in the order stated to properly interface with lsodes called in the main program of Listing A2.1. The composite derivative vector of the LHSs of eqs. (1.5) is calculated next and returned to lsodes.

- The dependent variable vectors are placed in three vectors to facilitate the programming of eqs. (1.5).

```
#
# One vector to three vectors
  u1=rep(0,n); u2=rep(0,n); u3=rep(0,n);
  u1t=rep(0,n);u2t=rep(0,n);u3t=rep(0,n);
  for(i  in 1:n){
    u1[i]=u[i];
    u2[i]=u[i+n];
    u3[i]=u[i+n2]
  }
```

Vectors are also defined for the LHS derivatives in *t* of eqs. (1.5).

- The first spatial derivatives $\frac{\partial u_1}{\partial x}, \frac{\partial u_2}{\partial x}, \frac{\partial u_3}{\partial x}$, are computed with splinefun (which is a utility in the basic R and therefore does not require an access statement in the main program of Listing A2.1).

```
#
# u1x
  tablex=splinefun(x,u1);
  u1x=tablex(x,deriv=1);
```

```
#
# u2x
  tablex=splinefun(x,u2);
  u2x=tablex(x,deriv=1);
#
# u3x
  tablex=splinefun(x,u3);
  u3x=tablex(x,deriv=1);
```

`deriv=1` specifies a first derivative.

- Homogeneous Neumann BCs are applied at $x = x_l, x_u$.

$$\frac{\partial u_1(x = x_l, t)}{\partial x} = \frac{\partial u_2(x = x_l, t)}{\partial x} = \frac{\partial u_3(x = x_l, t)}{\partial x} = 0$$

(A2.1a,b,c)

$$\frac{\partial u_1(x = x_u, t)}{\partial x} = \frac{\partial u_2(x = x_u, t)}{\partial x} = \frac{\partial u_3(x = x_u, t)}{\partial x} = 0$$

(A2.1d,e,f)

```
#
# BCs
  u1x[1]=0;u1x[n]=0;
  u2x[1]=0;u2x[n]=0;
  u3x[1]=0;u3x[n]=0;
```

Subscripts `1,n` correspond to $x = x_l, x_u$.

- The second spatial derivatives $\frac{\partial^2 u_1}{\partial x^2}, \frac{\partial^2 u_2}{\partial x^2}, \frac{\partial^2 u_3}{\partial x^2}$ are computed by differentiating the corresponding first derivatives (stagewise differentiation).

```
#
# u1xx
  tablexx=splinefun(x,u1x);
  u1xx=tablexx(x,deriv=1);
#
# u2xx
```

```
    tablexx=splinefun(x,u2x);
    u2xx=tablexx(x,deriv=1);
#
# u3xx
    tablexx=splinefun(x,u3x);
    u3xx=tablexx(x,deriv=1);
```

Note the use of `deriv=1`.

- Eqs. (1.5) are programmed over the interval $x_l \leq x \leq x_u$.

```
#
# u1t(x,t)
    for(i in 1:n){
      u1t[i]=a11*u1[i]+a12*u2[i]+a13*u3[i]+
            D1*u1xx[i];
    }
#
# u2t(x,t)
    for(i in 1:n){
      u2t[i]=a21*u1[i]+a22*u2[i]+a23*u3[i]+
            D2*u2xx[i];
    }
#
# u3t(x,t)
    for(i in 1:n){
      u3t[i]=a31*u1[i]+a32*u2[i]+a33*u3[i]+
            D3*u3xx[i];
    }
```

The result is the set of MOL/ODEs approximating eqs. (1.5). `u1xx`, `u2xx`, `u3xx` include a division by Δx^2 so this division is not included in the main program of Listing A2.1 (factors `D1dx2,D2dx2,D3dx2` are not defined in Listing A2.1 as they are in Listing 1.1).

- The three vectors u1t,u2t,u3t are placed in a single vector u of length $3(51) = 153$ for return to lsodes.

```
#
# Three vectors to one vector
  ut=rep(0,n3);
  for(i in 1:n){
    ut[i]    =u1t[i];
    ut[i+n]  =u2t[i];
    ut[i+n2]=u3t[i];
  }
```

- The counter for the calls to pde_A2 is incremented and its value is returned to the calling program (of Listing A2.1) with the <<- operator.

```
#
# Increment calls to pde_A2
  ncall <<- ncall+1;
```

- The derivative vector (LHSs of eqs. (1.5)) is returned to lsodes which requires a list. c is the R vector utility. The combination of return, list, c gives lsodes (the ODE integrator called in the main program of Listing A2.1) the required derivative vector for the next step along the solution.

```
#
# Return derivative vector
  return(list(c(ut)));
  }
```

The final } concludes pde_A2.

The numerical and graphical (plotted) output is considered next.

(A2.2) Model Output

Abbreviated numerical output is given in Table A1.3 of Appendix A1 (for comparison with the FD solutions) and is not repeated here to conserve space. Also, the graphical output is the same as in Figs. A1.1a,b,c of Appendix A1 and is not repeated here.

(A2.3) Summary and Conclusions

The MOL numerical integration of eqs. (1.5) with homogeneous Neumann BCs is straightforward. Also, the agreement of the FD and spline solutions in Table A1.3 implies that the spline collocation in `pde_A2` is valid. Variations of the PDE model, eqs. (1.5) with homogeneous Neumann BCs, could be accomplished such as Dirichlet, Robin BCs, nonlinear terms in the PDEs, addition of PDEs for more than three components. Also, computing and displaying the RHS and LHS terms of eqs. (1.5) would give additional insight into features of the solutions of these equations. These terms are readily available from the computed solutions u_1, u_2, u_3.

In summary, splines provide an alternative MOL algorithm (in addition to FD) that is convenient to use with the availability of `splinefun` in R. Splines generally have continuity (smoothness) properties that are not part of FDs, so the use of splines is recommended as an effective approach to MOL analysis of PDEs.

Index

Printed in the United States
by Bookmasters

Printed in the United States
By Bookmasters